PRACTICAL ORGANIC CHEMISTRY FOR THE BEGINNERS

PRACTICAL ORGANIC CHEMISTRY FOR THE BEGINNERS

Tanya Das
MSc (Jadavpur University), PhD (Pune University)
Reader in Chemistry, DA College, Garia

Amalendu Banerjee MSc, D.Phil.
Professor in Chemistry (Retired), Jadavpur University, Kolkata
Ex-Hon'y Lecturer, R K M R College, Narendrapur
Ex-Hon'y Teacher, Subhasini Balika Sikshalaya, Harinavi

New Central Book Agency (P) Ltd
LONDON
HYDERABAD ERNAKULAM BHUBANESWAR
DELHI KOLKATA PUNE

NCBA

REGD OFFICE
8/1 Chintamoni Das Lane, Kolkata 700 009, India
email: ncbapvtltd@eth.net

OVERSEAS
NCBA (UK) Ltd, 149 Park Avenue North, Northampton, NN3 2HY, UK
email: ncbauk@yahoo.com

EXPORT
NCBA Exports Pvt. Ltd, 212 Shahpur Jat, New Delhi 110 049
email: ncbaexp@ncbapvtltd.com

BRANCHES

208B Peacock Lane, Shahpur Jat
New Delhi 110 049
email: ncbadel@ncbapvtltd.com

House No. 3-1-315, 1st Floor
Nimboliadda, Kachiguda, **Hyderabad** 500 027
email: ncbahydb@ncbapvtltd.com

Shop Nos. 3 & 4, Vinayak Towers
681/B Budhwar Peth, Appa Balwant Chowk
Pune 411 002
email: ncbapune@ncbapvtltd.com

GSS Shopping Complex, 1st Floor
Opposite College Post Office, Convent Road
Ernakulam 682 035
email: ncbaernk@ncbapvtltd.com

Shop No. 15, Bharati Towers, Ground Floor
Forest Park, **Bhubaneswar** 751 001
email: ncbabub@ncbapvtltd.com

PRACTICAL ORGANIC CHEMISTRY FOR THE BEGINNERS

First Published: January 2011
Reprinted: February 2012

PUBLISHER AND TYPESETTER
New Central Book Agency (P) Ltd
8/1 Chintamoni Das Lane, Kolkata 700 009

PRINTER
New Central Book Agency (P) Ltd
Web-Offset Division, Dhulagarh, Sankrail, Howrah

COVER DESIGNER
Soumen Paul

COVER PRINTER
Biswanath Studio, Kolkata

ISBN 978 81 7381 651 2

Price ₹ 235.00

Dedicated to

Late Krishna Kumar Banerjee
and
Late Annapurna Banerjee
My Parents

Amalendu Banerjee

Dedicated to

Dr. Santa Banerjee
My Mother

Tanya Das

Theory without Practice and
Practice without Theory are both useless

CONTENTS

Preface

Chemistry is an experimental science where theory and experiment go hand in hand. So training in Practical Chemistry is essential for proper advancement of learning with equal importance to make the study of Organic Chemistry attractive and meaningful.

This book is written with a view to providing a whole-length Practical Organic Chemistry treatise to the graduate and postgraduate students of various institutions of our country. With this objective many items apparently not directly related to the practice have been incorporated here—hoping that these information would be helpful to the students and professionals equally.

This book is not much descriptive, rather, it is more of instructive type. It directs the stepwise procedure of the experiments in a how-to-do-it fashion with relevant caution, apposite notes, and indicating when the presence of the supervisors (teachers or teaching assistants) should be essential during the special or unusual type of apparatus setting and experiment at the initial stage.

It has been anticipated that the students using the book are already acquainted with the normal techniques of glass tube bending, cutting and fire-polishing of glass rod, jet drawing and preparation of long stem dropper, cork and rubber stopper boring, setting up a desiccator and the names of various apparatus and appliances used in chemical laboratories. So the directions in these regards and sketches of these articles have been omitted in this book. In many laboratories, the classical wash-bottle has been replaced by non-breakable and inexpensive plastic squeeze bottles of various sizes and different capacity. However, the use of "Calcutta Wash Bottle" has been proposed here from the sense of belonging to this great city of learning.

A problem of proper conducting of the practical classes arises from the increasingly large number of students in the graduate and postgraduate levels while the number of supervisors (teachers and teaching assistants) is not being increased proportionally. Hence, there is an increasing demand of self-explanatory books by the students who have to work independently (and they like it—and they should do so), unless the nature of the experiment demands the presence of a supervisor by the side of the students. Although such books act as "pseudo-supervisors", unfortunately no book will answer the uncommon and pertinent questions by the students which a supervisor does—so they are always needed by the students.

The contents of the book have been distributed in three chapters. The experiments included in this book have been mostly repeated in our laboratories very satisfactorily, and a few recent findings by us have been incorporated here to give the book an uptodate character. Chapter 1 deals with the preliminary suggestions on preparedness of the workers and getting acquainted with the basic techniques like solvent extraction, soxhlation, crystallisation, distillation, chromatography, synthesis, analysis and similar main aspects in Practical Organic Chemistry. This is followed by the preparations of several (nearly fifty representatives) common organic compounds in the macro-scale so that they could be utilised in subsequent experiments. Micro-scale derivative preparation of the unknown sample is also required of the student, so the test tube preparation of derivatives, their purification (crystallisation, centrifuge, chromatography, etc.) and identification has been given (about fifty).

Nearly fifteen well known "name reactions" have been furnished to make the students familiar with the textbook descriptions. It is anticipated that doing some "project works" will help the students in research work in future, hence a group of very simple project works has been listed here, the substrate had been already prepared by the students. To facilitate a quick reference of the common transformations of benzene and its derivatives a chart has been provided. As a very important content of the practical book, a list of laboratory reagents and solvents have been added in this part, together with the common catalysts.

The postgraduate students are expected to cultivate the habit of consulting chemical journals. To facilitate easy understanding of scientific communications some important terms, signs and symbols, and the methods of calculation of the percentage yield of the products have been added. Moreover, to help the students during write-up of their findings (classwork or research) a preliminary style guide—distinction between similar sounding pairs (a short list) in English, knowledge of common abbreviations, correct spelling of many English words (generally spelt wrongly by the students of our country)—has been provided. It is hoped that these additional information will infinitely improve the quality of the advanced students. In Chapter 1, the list of apparatus, diagrams, reagents and solvents required for individual preparation—and the probable mechanism of the reactions—have been omitted with the anticipation that the students will scan the given procedure and make the list all by themselves.

In Chapter 2, a short summary of the stepwise procedure for identification of the organic compounds has been given with later elaboration, explanation and equations. A special attention has been paid to the detection of various functional groups. An elementary idea of searching for (and also prediction of) λ_{max}, ν_{max}, τ values and fragmentation pattern has been given here under the heading of UV, IR, NMR, and mass spectroscopy, respectively.

Chapter 3 consists of the outline (basic principles) of the analyses of C, H, N, Hals, S and P, followed by determination of molecular formula through elemental, empirical and molecular weight determinations (if required). The importance of calculation of unsaturation number (also called degree of unsaturation) has been highlighted and eudiometric experiments to find out the molecular formula of hydrocarbons (in pure state or in mixture) have been included. The routine analyses of glucose, sucrose (pure or in combination with glucose), acetone, phenol, aniline, nitrogen (Kjeldahl's method), methoxyl and ethoxyl groups have been incorporated. The method of determining methoxyl and ethoxyl groups in a mixture, and the use of vanadate on perchloric acid in determining glucose and sucrose have been elaborated. A data bank containing some measurements has also been added for ready reference.

We believe that both understanding and conceptual power of the students could be infinitely improved through regular discussion on classroom analysis of samples, seminar talks, and project works. These topics must carry a sizeable percentage of the full marks of the practical examination. This will certainly reduce the strain and anxiety of uncertainty and luck factor of the examinee associated with the one or two days of examination. If it is not possible to attend all of the aforementioned topics a "grand viva" should be arranged for each student well in advance of the examination in the presence of several examiners and the average mark may be added to the credit of the examinee.

In fine we like to express our heartfelt gratitude to those who helped us ungrudgingly with their assistance, suggestions and constant encouragement during the preparation of this book. Special mention must be made of Late Professor Dr. Pratul Chandra Rakshit, the revered teacher of one of us (A.B.). We are also grateful to Sri Niharendu Banerjee, Dr. (Mrs.) Santa Banerjee, Sri Soumyabrata Das and Mrs. Reba Das for their constant interest in the writing of this book. Thanks are due to Dr. Haraprasad Samaddar, Dr. Rina Ghosh (Jadavpur University), Dr. M. M. Adak (Asutosh College), Dr. (Mrs.) Anuva Putatunda (Lady Brabourne College), Dr. Sachchidananda Dutt and Dr. G. C. Banerjee (both retired teachers of Jadavpur University), Dr. Santanu Banerjee (Nut Behari School), Dr. Arogya V. Saha and the learned teachers of R. K. Mission Residential College, Narendrapur, the reputed collagues of D.A. College, Garia, of one of us (T.D.) and Sri Uttam Guha (Publishing Orbit) for their helpful criticism and inspiration. One of the authors (A.B.) is extremely grateful to Dr. K. B. Baksi, MD (Cal), Dip. Card (Cal.), F.A.C.C. (USA), F.C.C.P. (USA) for his keen interest and treatment to keep him (A.B.) fit and healthy during the making of the book.

We also thank Shri Amitabha Sen, Director of New Central Book Agency (P) Ltd., for his interest in publishing the book; and the NCBA production team for their unstinted and rigorous toil in bringing it to fruition.

It is anticipated that in this book some undetected errors and flaws exist, which would be corrected as soon as they would be pointed out by the esteemed teachers and interested students.

We will be really glad if this book receives appreciation of our kind readers.

January 2011

TANYA DAS
AMALENDU BANERJEE

Preparative Practice and Other Information

1

1.1 INTRODUCTION

Chemistry is an experimental science. Experiments lend support to the theory and they are complementary to each other. Good experimentalists are always appreciated and often they are referred to as "workers".

Generally, the laboratory work consists of three distinct parts, viz., experiment, observation and inference (i.e. conclusion). The inference drawn may be wrong due to several reasons, but wrong observation is offensive and annoying. So extreme care should be exercised during observation of the progress of the reaction.

A good worker should enjoy good health, eyes checked for colour blindness or poor vision, nose checked for smelling subtle odour. Other sensory organs are not very important, since testing the sample by tasting is strictly forbidden (the sample may be highly poisonous), such is the case with ears and skin. The workers are not encouraged to touch the unknown samples which might be toxic, carcinogenic and deadly poisonous.

The workers should have excellent emotional stability. They must be able to work in a group of coworkers, even in foreign environment, and withstand the occasional monotonous, bone-crushing and dangerously hard work in the laboratory resulting in frustration due to failures.

At the beginning the workers should note the following suggestions:

(a) Be fearless but be extremely careful.

(b) Know all the hazardous, inflammable, toxic, corrosive, etc., chemicals and reagents that you handle. Feel free to see your supervisors who will be very happy to show you the various steps to be done, e.g. setting up of the apparatus, follow-up of the reaction, work-up of the reaction mixture, isolation and preservation of the desired product, etc. at the beginning. A good supervisor is always desirable as he can make another good supervisor.

(c) Know the locations of the fire-extinguisher, fire-escape, first-aid box, telephones, exhaust fans and fume-chambers. Keep your head cool if unfortunately fire breaks. Learn to operate fire-extinguishers in case of emergency.

(d) Keep your personal eye-protector, safety-razor blade, match box, a length of strong threads and a clean towel or a piece of cloth. The wet or sticky hands of the worker may pose lots of problems while working in the laboratory.

(e) The following "Dont's" are to be obeyed:

- (i) Don't take the laboratory as a playground. Be careful, a small mistake may cause great harm.
- (ii) Don't pollute the laboratory environment by preparing large quantity of obnoxious smelling and harmful compound outside the fume-chamber.
- (iii) Don't block the free flow of water anywhere in the laboratory and make the floor slippery.
- (iv) Don't keep food or drink in a refrigerator where chemicals and solvents are stored.
- (v) Don't let the gas leak from Bunsen burners when not in use.
- (vi) Don't be too adventurous to perform an exploratory experiment without proper precaution and a companion in the laboratory if the supervisor is not available.
- (vii) And finally, don't smoke and starve yourself.

1.2 SEPARATION OF THE COMPONENTS OF A MIXTURE

The sample may be isolated from the natural products by solvent extraction at room temperature or by soxhlation (a special method of continuous extraction of the sample by solvents under reflux in a special apparatus). Also the sample may be prepared in the laboratory from various ingredients to train students and examine them.

The organic compounds are generally of four types, viz., A_1, A_2, B and I which are elaborated in the following:

(a) Strong organic acids (but certainly weaker than the mineral acids) react with $NaHCO_3$ in aqueous solution to furnish water soluble salts of organic acids with evolution of CO_2 gas. Such organic acids belong to the type A_1 (e.g. carboxylic acids, etc.) which turn blue litmus red (moist) but do not affect congo-red which turns blue with mineral acids. So from the salts of organic acids complete liberation of the acids will be effected by minerals acids using congo-red as the indicator which turns blue as soon as the inorganic acid becomes excess. Blue litmus is unsuitable for this purpose.

(b) Weak organic acids (A_2) do not react with weak base ($NaHCO_3$), but dissolve in strong alkali (NaOH or KOH) forming salts which regenerates original A_2 by acidification or better by passing CO_2 through the salts of A_2 in aqueous solution (e.g. of phenolic compounds).

(c) Organic basic compounds, B (mainly amino) make water soluble salts with mineral acids. These salts liberate the original organic bases on treatment with aqueous alkali which is stronger than the organic bases (B).

(d) Indifferent or neutral compounds [I] remain unaffected by mineral acids or alkalis. So after the stepwise removal of A_1, A_2 and B from the mixtures, [I] remains in the organic solvent. Thus the proposed plan of separation is shown in the following scheme [1.3].

1.3 THE SCHEME OF SEPARATION OF THE COMPONENTS OF A MIXTURE

(a) The sample is prepared by mixing A_1, A_2, B and I, in any combination and proportions.

(b) The sample is tested for complete and partial solubility in various solvents viz., benzene, petroleum ethers (40–60° and 60–80°), chloroform, dichloromethane, carbon tetrachloride, methanol (toxic) or ethanol, acetic acid and water. When the mixture is completely soluble in water, the separation becomes difficult but not impossible. In most of the cases, the sample completely dissolves in organic solvents like C_6H_6, Pet. ether, $CHCl_3$, $CHCl_2$, CCl_4, ether and ether-benzene mixture (i.e. $C_6H_6 + Et_2O$, 1:1, v/v) which are insoluble in water.

c) The weighed sample (mixture of A_1, A_2, B and I, $5{\cdot}x - 10{\cdot}y$ *gm.* (*x* and *y* are known) is dissolved in such appropriate solvent (selected on the basis of cost, volatility, density, boiling point, safety in handling, etc.) and transferred in a separatory funnel followed by stepwise extraction with the aqueous solutions of inorganic reagents as shown below.

(d)

Step (i) →	*Step (ii)* →	*Step (iii)* →	*Step (iv)* →	*Step (v)*
The mixture is extracted with dist, H_2O (50 ml × 3). The combined extract is evaporated to dryness. The residue (if any) is weighed and identified.	The solution is extracted with sat. $NaHCO_3$ soln. (5%, 50 ml × 3). Excess of $NaHCO_3$ must be ensured [–A_1]	The solution is extracted with aq. NaOH soln. (8% 30 ml × 3). Excess of alkali ensured (red litmus) [–A_2].	Washed with water to remove alkali from the soln. Then extracted with dil.HCl (6N, 30 ml × 3). Excess of acid is ensured [–B].	The solvent washed free from HCl with H_2O (20 ml × 3) and solvent removed. The residue dried, weighed and identified [I].
Water soluble part is removed from the mixture, marked [S].	The combined extract taken in a beaker on acidification with conc., HCl, generates A_1. This is isolated, dried, weighed and identified [A_1]	The combined extract on acidification with dil. HCl (6N) liberates A_2 which is collected, dried, weighed and identified [A_2].	The combined extract basified with dil. NaOH and the liberated product isolated, dried, weighed and identified [B].	

(e) Separation of components belonging to the same group (like A_1, A_2, B and I): The initial weight of the mixture taken and the weight of the isolated material, at any step will indicate the nature of the mixture. Thus Wg of the mixture yielding very nearly Wg of A_1 will indicate the mixture of two (or more) strong organic acids. In such case, the yields of A_2, B and I should be almost nil. This type of mixtures are to be separated by special techniques described afterwards [vide 1.6, B (a), ii].

1.4 SOLVENT EXTRACTION

(a) The regenerated component of the mixture [vide 1.3(d)] in most of the cases is collected by solvent extraction. The liberated compound is dissolved in appropriate solvent and transferred to a separatory funnel. An Erlenmeyer flask (100 ml) fitted with a bored cork through which a small (ca 6–7 cm) glass jet is fixed and called "Calcutta wash bottle" which contains the same solvent used for the extraction. This is used to wash all the apparatus to minimise "work-up" loss.

The separatory funnel is closed with a rubber stopper, initially shaken very carefully with occasional release of the internal pressure, then vigorously and allowed to settle by leaving it on an iron ring. The contact of iron with glass is undesirable, so the ring is wound with a piece of cloth or string.

The separated layer is removed and collected in an Erlenmeyer flask. The aqueous layer is again extracted (if x ml of solvent is used each time and repeated for n times, then this will be shown as x ml × n), and the combined organic layer is washed with water (or with brine if the desired material is soluble in water), dried over anhydrous sodium sulphate, filtered through cotton and solvent removed. The mode of removal of the solvent depends on the boiling point and characteristic properties of the solvent. In many cases, the solvent may be recovered and reused. This whole procedure of solvent extraction—washing, drying and removal of the solvent to isolate the crude product—is often abbreviated as "work-up" of the reaction mixture.

(b) Commercial sodium sulphate is strongly heated in an iron pot to remove water of crystallization (ca one hour heating) and the hot anhydrous material is stored in a wide-mouth stout bottle fitted with a rubber stopper. When cold this is used as anhydrous sodium sulphate.

(c) If the solvent is heavier than water, two separatory funnels should be used simultaneously, so that the work-up becomes more easy. The contents of the separatory funnel should not exceed three-fourths of the capacity of the funnel.

The residue obtained after the removal of the solvent is further dried under vacuum or in a chamber heated to suitable temperature. The weight of the dried product is used to determine the percentage composition of the given mixture.

(d) In the separatory funnel the layers will separate as shown below:

Separatory funnel

Upper layer	H_2O	H_2O	H_2O	H_2O	H_2O	C_6H_6/Tol.	Et_2O	C_6H_6/Et_2O	
Lower layer heavier	Ph-NO_2	CH_3I	CH_2Cl_2	$CHCl_3$	CCl_4	H_2O	H_2O	H_2O	
Density (bp)	1.198 (210)	2.28 (42)	1.33 (40)	1,5 (61)	1.6 (77)	0.87/0.86 (80/110)	0.71 (34)	–	Organic solvent bp = °C.

* After each completion of extraction, the stop-cock of the separatory funnel should be washed with water, wiped dry on cloth, greased well and secured by a rubber band or a string.
* The solvent should never be removed on a direct flame, low boiling solvents are removed from the steam bath and high boiling (bp > 100°C) solvents are removed from the steam bath under reduced pressure (rotary evaporator).
* If excess of alcohol is present in the reaction mixture (as in esterification) then a portion of the alcohol is removed from a steam bath before solvent extraction of the desired product. Excess of alcohol, if present, will pose serious problem during separation of the layers. Also in alkali extraction annoying emulsion may be formed which can be controlled by adding little electrolyte [NaCl] in the separatory funnel:

(e) Various organic solvents and reagents are dried (with/over):

(i) $CHCl_3$ (anhydrous $CaCl_2$ or anhydrous K_2CO_3).
(ii) MeOH or EtOH (stepwise, CaO → CaH_2 → Mg) → absolute alcohol.
(iii) Ether (diethyl) (fused $CaCl_2$).
(iv) Acetaldehyde (anhydrous Na_2SO_4).
(v) Acetone (anhydrous K_2CO_3, obtained by heating commercial K_2CO_3).
(vi) Liquid organic acids (anhydrous Na_2SO_4 or $MgSO_4$ or $CaSO_4$).
(vii) Benzene or aromatic hydrocarbons (liquid) (Na or fused $CaCl_2$).
(viii) Ethyl acetate (anhydrous $CaCl_2$ or K_2CO_3).
(ix) Phenol (anhydrous, K_2CO_3).
(x) Aniline or any amine (liquid) (solid KOH balls).
(xi) Nitrobenzene (anhydrous, $CaCl_2$).
(xii) Alkyl halide (liquid) (anhydrous $CaCl_2$ or Na_2SO_4 or $MgSO_4$ or $CaSO_4$).
(xiii) HCO_2H (boric anhydride).

$MgSO_4$ is fine powder. It is difficult to separate by filtration. This is an excellent dehydrating agent.

1.5 PURIFICATION OF ORGANIC COMPOUNDS

This is a very important task of the organic chemists as it involves good manipulative skill (sometimes less than ten milligram of the sample is available), patience and practice (and also a little bit of luck). The following are the classical methods of purification (samples for elemental analysis are termed analytically pure). Some more instrumental methods are used nowadays. These are superior to the classical methods, as they do not involve the personal skill. These methods are ultracentrifugation, electrophoreses, zone melting technique, G.L.C. dialysis, diffusion, molecular distillation, etc.

The classical methods are:

(i) **Crystallisation** The mp of impure solid sample is always lower than that of pure sample. So the sample is crystallised repeatedly until the melting points for three consecutive crystallisations remain constant.

At the starting point, the solubility of the sample is tested for organic solvents (pure and bp range preferably between 60 and 80°C) are better than water as the latter required higher temperature and longer time for complete removal to render the analytical sample. A small amount of the sample is taken in a clean test tube (t.t.), few drops of solvent are added to it and warmed on a steam bath, dropwise addition of the solvent is continued till the sample dissolves completely, shaking the t.t. should be continued all the while to avoid bumping. On cooling crystals should appear, if not other solvents are to be attempted.

After the choice of the proper solvent (the sample neither be too much soluble nor too much insoluble) the sample is taken in the clean appropriate flask (Erlenmeyer flask, never in a beaker unless water is the solvent) and dissolved in the solvent by heating on a steam bath then filtered. A little excess of the solvent is added to it to prevent the crystal formation during filtration, which should be done on a filter paper. The filtrate must be optically pure (no suspended impurities) and collected in a clean Erlenmeyer flask, then a few boilizers are added to it (the capillary tube sealed at the middle is the best boilizer as it furnishes the "seed" of the sample on evaporation of the solvent) and concentrated on a steam bath. The crystals deposited on cooling (sometimes "seeding" is necessary) are filtered (this time may be under the suction of water aspirator using small "Euchner" or "Hirsch" funnel), washed with a small amount of cold solvent, dried and melting point noted. When the melting point becomes constant by several crystallisations, the sample is dried in a heating chamber under vacuum, analyzed and preserved after proper labelling.

All the apparatus needed for analytical sample preparation should be heated with concentrated nitric acid on a steam bath for several hours, then washed with doubly distilled water and carefully dried in a heating chamber preferably electrical. The boilizer is prepared by heating small pieces of porcelain in the above procedure.

Very small amount of the crystals (less than 5 mg) may be separated from the solvent by centrifugation or with the help of a dropper, the tip of which is bent and drawn into a jet. If the crystals do not appear within reasonable time, the seeding (vide supra) will initiate the crystallisation. If the solution is left undisturbed, large crystals will be formed but small crystals are more suitable for analysis, so to get such sample the solution is gently disturbed by vibration like tapping the platform on which the flask rests.

(ii) **Distillations (simple, fractional and steam distillation)** This technique is easier than fractional crystallisation. Claisen flask is better than ordinary distillation flask. The fractionating column Vigreūx is suitable for better purification of liquid samples. The impurities elevate the boiling point. For analytical sample collection the receiver should be analytically clean. At the starting point boilizer must be added, even when the liquid was boiled then cooled before hand. Boilizer should not be added to the very hot (yet not boiling) liquid. Distillation under reduced pressure requires capillary dipped into the liquid. Large volume of liquid when distilled may be carried out over steam bath, oil bath or sand bath depending on the boiling point. Steam distillation is carried out by passing steam through the boiling liquid, here no boilizer is required.

(iii) Evaporative distillation (sublimation) This is suitable for making small amount (not enough for distillation) of impure liquid into analytically pure sample. This is done in a sublimation tube (hard glass, about 7″ long and 1.5″ in diameter with a constriction towards the closed end to make a well) which contains the impure sample. A little of glass wool is added to it as a plug to prevent spirting, then inserted in an electrically heated chamber and subjected to evaporation under vacuum. The distillate is slowly deposited in the well of the sublimation tube. This is collected and analysed.

(iv) Chromatography This is a very useful and widely used technique to prepare pure sample. Various types of chromatography are in use of which the column chromatography is most common. This technique consists of making a column of absorbent (usually alumina, neutral, acid washed or basic) in a tube (column height at least 7 times of the diameter of the tube and absorbent at least 30 times by weight of the sample to be purified). This is done by taking the appropriate organic solvent in the tube fitted with a stop-cock, and slowly adding the measured absorbent by gently tapping the column. There should be uniform deposition of the absorbent and no air bubble should be trapped in. The impure sample is dissolved in the same solvent, added over the column and gradually washed down by the same solvent (this is called elution). The solvent is removed and the residue examined. If needed the polarity of the solvent gradually increased to isolate different classes of compounds in the pure state.

1.6 EXPERIMENTS FOR TRAINING OF THE BEGINNERS

The following experiments may be helpful during training of the organic chemists:

(A) Oxidation

(a) By $KMnO_4$ Mol. wt. 158; Eq. wt. 31.6 (secondary standard solution), soluble in water, strong oxidant (potential 1.54 volt), oxidises under neutral, acidic, and alkaline conditions, soluble also in 18-crown-6 solvent. Generally, the substrate is finally oxidised to the carboxylic acid:

General procedure of oxidation:

(i) In a beaker (250 ml) $KMnO_4$ (2.5 g) is dissolved in water (dist, 40 ml).

(ii) In an Erlenmeyer flask (250 ml) the substrate (ca 1 g) is taken and dil. NaOH (1 ml, 6N) is added to it. A few boilizers are added to the mixture, then a reflux condenser is fitted to the flask which is set on an wire gauze and heated.

(iii) The freshly prepared (i above) $KMnO_4$ solution is added portion-wise (10 ml × 4) from the top of the condenser (the water inlet will be always at the lower level) to the hot mixture which is gradually heated under reflux. Each portion of purple $KMnO_4$ solution becomes almost colourless when the next portion is added. If needed a little more $KMnO_4$ solution should be added to complete the reaction (purple colour is not discharged any more).

(iv) After the completion of the addition, the reaction mixture is boiled for nearly 1 hr. Care should be exercised to control the violent jerking of the reaction mixture (R.M.) due to formation of brown MnO_2 as ppt. during this period.

(v) Then the R.M. is treated with HCl (conc) until acidic (congo-red) and to the hot R.M. is added $NaHSO_3$ (or Na_2SO_3), little at a time with vigorous stirring until the solution becomes clear and colourless. On cooling in ice the desired oxidation product will crystallise.

(vi) This precipitated product is filtered on a Buchner funnel under suction (water aspirator), washed with cold H_2O until free from mineral acid (congo-red indicator, not blue litmus) dried and weighed (percentage yield is calculated), m.p. noted, and mmp determined with authentic sample, if available.

(vii) If the desired product is liquid or soluble in water, then the solution is treated with a little NaCl, stirred well to dissolve the NaCl, and extracted with ether (thrice with 50 ml ether each time = 50 ml × 3). The combined ethereal soln. is washed with brine (saturated solution of NaCl in H_2O), then with cold water (5 ml), dried over anhydrous Na_2SO_4, and solvent removed over steam bath. The residue, thus obtained, is distilled (if needed under reduced pressure) to furnish the desired product which is weighed, percentage yield calculated, and confirmed through m.p. and m.m.p of its derivatives.

(viii) It the reaction product is a mixture of unreacted starting material and carboxylic acid then the combined ethereal solution (vide vii) is extracted several times with saturated aqueous solution of $NaHCO_3$ (test with red litmus). The combined alkaline extract is acidified with cold and conc. HCl (caution, there would be vigorous gas evolution) and again extracted with ether (sometimes a mixture of ether and benzene (1:1, v/v) is used to prevent loss of ether particularly during summer, although benzene is hazardous to health) for several times. The combined organic layer is washed once with brine, then twice with cold water (5 ml × 2) dried over Na_2SO_4 and solvent removed. The purified acidic product, thus obtained, is collected by the afore-mentioned procedure (vide vii).

(ix) The initial ethereal solution (viii) is washed free from alkali (H_2O, 10 ml × 2), dried (Na_2SO_4), solvent removed over steam bath and the residue (unreacted starting material is substrate) collected in the appropriate way (crystallisation or distillation) to calculate the material balance which indicates the loss of substrate during the reaction and work-up.

NOTES:

- The distillation (or boiling) should never be done in a closed vessel. A $CaCl_2$ "guard tube" fitted at the "exit point" would make the reaction or distillation carried out under anhydrous condition, when the reaction is done strictly under nitrogen or hydrogen atmosphere then the exit point should be dipped under mercury.

- To the very hot (but not boiling) liquid should never be added the boilizer. This may lead to severe accident.
- If the reaction mixture contained acetic acid, the extraction should be done with chloroform instead of ether which dissolves HOAc to the greater extent.
- In brine the organic acids are less soluble than in water.
- If the reaction mixture contained lots of alcohol (i.e. methanol or ethanol, as in esterification), a portion of the alcohol should be removed before solvent extraction, otherwise, the separation of the aqueous and organic layers in the separating funnel becomes very difficult (addition of NaCl may not be helpful).
- The condenser should not be held too tightly by the clamp. It should rotate with the help of the fingers. This will protect the condenser from cracking. Also it should stand on the wire gauze in a slightly slanting position away from the worker.
- While generating organic acids from their sodium salt in alkaline solution by acidification with mineral acids (HCl, H_2SO_4 etc.) congo-red indicator should be used which turns blue when organic acid generation is completed and inorganic acid becomes excess. Blue litmus indicator cannot serve this purpose as it is affected by the liberated organic acids while congo-red is affected only by mineral acids.
- Fresh $NaHSO_3$ (or Na_2SO_3) should be used to dissolve the MnO_2 formed, otherwise, large excess of the reagent has to be added and lots of elemental sulphur will be formed resulting in impure product.
- The same and pure solvent should be used for washing purpose of the product from which it has been crystallised.
- It is better to dissolve the MnO_2 rather than filtration under pressure as the solid MnO_2 would retain some of the desired product.
- In a beaker (ca 500 ml) water is boiled (this is the steam bath) and the product to be dried is taken on a watch glass which is placed over this steam bath. The m.p. of the product should be at least 15–20°C higher than 100°C. If the m.p. is below 100°C, then the sample may be dried under vacuum and at lower temperature. It may be dried also in an electric oven where the temperature is controlled at appropriate point.
- In many countries toluene (bp. 110°C) is used in place of benzene. Toluene may be removed under reduced pressure from a steam bath.
- Water insoluble substrate may be dissolved in 18-crown-6 and oxidation by $KMnO_4$ could be carried on as $KMnO_4$ is soluble in this solvent (but the solvent is very expensive).

Specific example: *Preparation of m-nitrobenzoic acid, m.p. 140°C, from m-nitrotoluene* In an Erlenmeyer flask (250 ml) fitted with a reflux condenser was taken *m*-nitrotoluene (1 g), $KMnO_4$ (2.5 g in 40 ml water) and NaOH solution (6N, 4-5 drops) and heated under reflux for one hour when the purple colour of $KMnO_4$ was discharged and brown precipitate of MnO_2 was obtained. To the hot solution, a small amount of $NaHSO_3$ was added followed by acidification with conc. HCl (congo-red) with stirring when the reaction mixture becomes colourless and clear. On cooling in ice, colourless crystals of *m*-nitrobenzoic acid were obtained which were collected by the usual process (yield 0.8 g, 60-65%, m.p. 138-139°C). If required it may be recrystallised from hot water.

(b) By $K_2Cr_2O_7$ and dil. H_2SO_4 $K_2Cr_2O_7$, mol.wt. 294, eq. wt. 49.3; moderately strong oxidant, redox potential 1.33 volts, generally used to oxidise 1° and 2° alcohols into oxo-compounds.

General procedure for oxidation

(i) In a 100 ml Claisen flask (a distillation flask with a side arm having a mercury well on the top of it where the thermometer is dipped to note the boiling point of the distillate) is taken the sample (1 g) and dil. H_2SO_4 (1 N, 20 ml) and $K_2Cr_2O_7$ (0.5 g). A few boilizers are added to the mixture and the flask is stoppered well. The side arm of the flask is connected to the inclined small condenser, the other end of it is introduced in a small receiver (25 ml).

(ii) The reaction mixture (R.M.) is gently boiled for 10 mins. taking care that no distillate is collected. During this period the initial yellow clour of the R.M. gradually changed into light green. Then the heat is increased to distil the desired product with water (total volume is about 10 ml) which is collected in the receiver.

(iii) This distillate (10 ml) is transferred into a separatory funnel (50 ml), extracted with ether (5 ml × 3), washed with H_2O (5 ml) and dried (Na_2SO_4). On removal of the solvent on a steam bath the desired product (0.5 g) is obtained as residue which may be identified through its derivatives. The yield is generally 50-60%. When larger amount of oxidised product is required to determine the b.p. and as substrate, the quantity of the starting materials and the corresponding size of the apparatus should be increased appropriately.

NOTES:

- Condensers are of two types, viz. water condenser (water flows from lower level inlet to the higher level outlet), and air condenser (sometimes a wide glass tube will serve the purpose).
- The free end will go inside the receiver which may be cooled in ice if needed. The receiver will never be stoppered during distillation but a $CaCl_2$ guard tube can be attached to the side tube of the receiver when distillation under anhydrous condition is essential.

Specific example: *Cyclohexanone from cyelohexanol* To the mixture of cyclohexanol (10 g), $K_2Cr_2O_7$ (5 g) and dil. H_2SO_4 (1N, 200 ml) contained in a Claisen flask (500 ml) was added a few boilizers and the mixture was slowly distilled using a water condenser. The distillate (100 ml) was collected, extracted with ether (50 ml × 3) washed with brine (10 ml), dried (Na_2SO_4) solvent removed (steam bath) to furnish a residue (approx. 9 g) which was distilled to collect the pure cyclohexanone (8.5 g, 80-85%) bp., 155-156°C; 2,4-DNP, mp 162°C.

For identification purpose the distillate from the original reaction mixture is directly collected into a solution of 2,4-dinitrophenylhydrazine sulphate when the yellow 2,4-DNP derivative is obtained. This is filtered under suction and chromatographed over Al_2O_3 (using benzene) to furnish pure 2,4-DNP, m.p. 162°C.

(c) By conc, HNO_3; Mol. wt. 63; density 1.42; very strong acid and oxidant; dangerous chemical; also used with conc. H_2SO_4 for nitration of aromatic compounds.

General procedure for oxidation:

(i) In a porcelain basin (100 ml) was taken the sample (1 g), heated on a steam bath and treated with conc. HNO_3 (added dropwise) when a vigorous reaction started. Two drops of HNO_3 should be added each time and the R.M. was stirred with a glass rod when dark-brown fumes evolved. When the reaction slowed down the next portion (2 drops) was added in the similar way until the reaction was completed (no brown fumes on addition of HNO_3).

(ii) The desired oxidised product was obtained as a yellow solid in the porcelain basin. This product was cooled in ice and washed with cold ether (2 × 15 ml) when the adhering HNO_3 was washed away and colourless solid was derived. This became dry immediately and collected, weighed and m.p. determined.

NOTES:

- To prevent environment pollution by obnoxious fumes this oxidation should be carried out in a fume-chamber or under good exhaust fans.
- The basin holding the oxidised product should be cooled in ice (to prevent loss of ether during washing) then washed with cold ether to remove the unreacted HNO_3 absorbed by the solid desired product. This HNO_3 imparts the pale yellow colour. The desired product may be slightly soluble in water. So it is advisable to use cold ether, when small quantity of product is to be washed, the dropper may be used to add the solvent, stir the product and remove the washings. In this way filtration of small quantity could be avoided.

Specific example: *Adipic acid from cyclohexanol (or cyclohexanone).* To the sample (1 g) contained in a porcelain basin (100 ml) and warmed on a steam bath was added portion-wise (2 drops at a time) conc. HNO_3 when vigorous reaction started with evaluation of brown fumes (use fume chamber). When the reaction was over (no more brown fumes even on addition of the oxidant) the basin was cooled in ice, the contents were washed with cold ether to remove the excess of HNO_3 (cause of pale yellow colour of the product) with the help of a dropper. The residue became dry at once, which was collected, weighed (0.9 g, 65-70%), m.p. 150°C.

(d) By $KBrO_3$ Mol. wt. 167; Eq. wt. 27.84; moderately strong oxidant; redox potential 1.42 volts, solvent glacial acetic acid.

General procedure

(i) In a round bottom flask (100 ml) fitted with an air condenser (12″-15″ long) was taken a mixture of gl. HOAc (15 ml) and substrate (1 g) to which was added powdered $KBrO_3$ (0.5 g) and slowly heated. Gradually, elemental Br_2 was liberated and the R.M. assumed orange red colour. The mixture was heated for 10 mins taking care that no solvent was lost, then cooled to room temperature (r.t.) and again $KBrO_3$ (0.2 g) was added to it and heated for 5 mins.

(ii) The aforementioned R.M. (in *i*) was cooled and poured onto water (100 ml) and extracted with $CHCl_3$ (20 ml × 3). The combined organic layer (ca 60 ml) was washed with H_2O (10 ml), excess of aqueous $NaHCO_3$ solution (preserved for iii), followed by water to free the $CHCl_3$ from $NaHCO_3$. On drying (Na_2SO_4) and removal of the solvent was obtained the neutral residue which was collected and identified in the usual way.

(iii) The aforementioned alkaline layer (ii) was cooled in ice and acidified with dil. H_2SO_4 (4 N) using congo-red indicator when the desired carboxylic acid was obtained as solid ppt. which was filtered under suction, washed (H_2O) dried, weighed and identified in the usual way.

(iv) If the desired acid was liquid or water soluble then the acidified solution (iii) was made saturated with solid NaCl (salting out) and extracted with ether (20 ml × 3). The combined ethereal layer was washed with cold water (5 ml) and dried (Na_2SO_4). On removal of the solvent over steam bath the residue obtained was collected, weighed and identified in the usual way.

(v) The carboxylic acids were identified through m.p., m.m.p. (if solid) and preparation of apposite derivatives.

NOTES:

- To prevent environment pollution, the open end of the air condenser was connected to a rubber tube, the other end of which is dipped under running water tap.
- Oxidation by $KBrO_3$ is a better procedure than using $K_2Cr_2O_7$ and $KMnO_4$ as Cr, Mn, etc. cause environment pollution.
- $CHCl_3$ does not extract HOAc from aqueous solution, hence $CHCl_3$ is the right choice as extracting solvent in this example.
- The saturated solution of $NaHCO_3$ contains 5% $NaHCO_3$. Only 10% excess of the calculated volume of $NaHCO_3$ would be sufficient for complete extraction of acidic material.
- If the desired product be neutral then the alkaline extract may not be preserved.
- This procedure is less time consuming and work-up process is more easy compared with many oxidising agents of comparable strength.

Specific examples

(i) *Conversion of 4-nitrobenzaldehyde into 4-nitrobenzoic acid in 80–85%* yield is done by the general procedure, m.p. 240–241°C (a good derivative of the substrate).

(ii) *Preparation of aromatic ketones and quinones by $KBrO_3$* In a round bottom flask (100 ml) were taken sample (1 g), $KBrO_3$ (1 g), and gl. HOAc (15 ml). An air condenser was attached to the flask which was heated slowly on a low flame for 20 mins. Then the RM was cooled and another portion of $KBrO_3$ (1 g) was added to it and boiled for further 15 mins taking precautions to remove evolved Br_2 under the water tap.

(iii) This R.M. was poured onto cold water (150 ml) and the solid quinone which was precipitated was filtered, washed, dried (over steam bath), weighed and identified in the usual way.

NOTES:

- This oxidation by $KBrO_3$/HOAc is not suitable for acyclic or alicyclic compounds because large amount of undesirable bromo-carbonyl compounds which are very irritating to the eyes are formed.
- It the starting aromatic compound is anthracene, then the R.M. is heated on a low flame for 10 mins, then cooled to r.t., fresh $KBrO_3$ is added, again heated for 10 mins, then cooled and poured onto water. The yellow ppt. of 9,10-anthraquinone is filtered under suction, washed with water, then with ethanol, dried and if required crystallised from benzene, m.p. 283–285°C.
- If the sample is phenanthrene then the heating period is increased to 25 mins for each of two instalments of addition of $KBrO_3$, thus making the total reaction period of 50 mins. Then the R.M. is poured onto water and the ppt. of 9,10-phenanthrenequinone (orange coloured) was filtered, washed with ethanol, dried and if required crystallised from benzene, m.p. 200–202°C.
- Diphenylmethane and benzilic acid furnish benzophenone (m.p. 48°C) by oxidation with $KBrO_3$/HOAc in the aforementioned method. 9-carboxy-9-hydroxyfluorene yields fluorenone (m.p. 83°C) by the $KBrO_3$/HOAc oxidation by this method.

(e) By $NaBO_3 \cdot 4H_2O$, sodium perborate tetrahydrate. It is readily available, stable and mild oxidant. It does not cause environment pollution.

General procedure

(i) In an Erlenmeyer flask (100 ml) was taken a mixture of sample (0.2 g – 0.5 g) and glacial HOAc (3 ml for each gram of the oxidant) and warmed to 95°C over a steam bath. To this hot mixture was added $NaBO_3 \cdot 4H_2O$ (1 g –2 g) with swirling and left on the steam bath for 30 mins with occasional swirling. This addition of the oxidant and heating for 30 mins was repeated thrice (total oxidant 4 g – 8 g, and time 2 hrs). The RM was cooled and poured onto cold water (150 ml) then extracted with $CHCl_3$ (20 ml × 3) (because HOAc was present). The combined organic extract was washed with water then $NaHCO_3$ (15 ml × 3, this is preserved), again H_2O (10 ml × 2), dried, and solvent removed to furnish the desired neutral oxidation product which was identified in the usual way.

(ii) If the desired oxidation product was acidic then the $NaHCO_3$ extract (in i) was cooled in ice and acidified (congo-red indicator) with conc. HCl. The liberated carboxy acid was extracted with chloroform or ether (20 ml × 3), and the combined organic layer was washed with water, dried (Na_2SO_4), and solvent removed. The residue was collected and identified.

NOTES

- In this process very pure desired products are obtained in excellent yields (80-100%).
- *Specific examples*

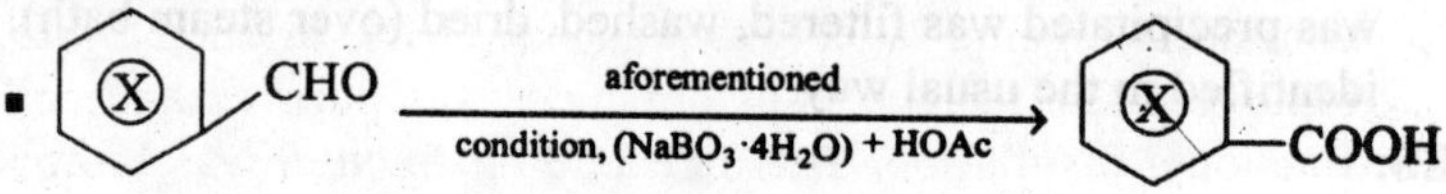

X = H, Cl, NO_2, OMe, and OH (*o*, *m* and *p* isomers)

- $Ph_2C(OH)(CO_2H) \xrightarrow[HOAc, \Delta]{NaBO_3 \cdot 4H_2O}$ Ph–CO–Ph (m.p. 48°C)
- Cyclohexene or cyclohexane-1,2-diol $\xrightarrow{Do}$ adipic acid (m.p. 150°C),
- 9,10-Phenanthrenequinone $\xrightarrow{Do}$ 2,2′-dicarboxydiphenyl (m.p. 229°C),
- $Ph_2C(OH)–C(OH)Ph_2 \xrightarrow{Do}$ benzophenone (m.p. 48°C),
- Ph–CH = CH–Ph $\xrightarrow{Do}$ 2Ph–CO_2H (m.p. 122°C),
- Ph– C ≡ C–Ph $\xrightarrow{Do}$ 2 Ph–CO_2 H (m.p. 122°C),

[B] Halogenation

(a) Bromination with $KBrO_3$, glacial HOAc and dil. H_2SO_4(8 N)

(i) *For substituted benzenes:* To the mixture of sample (10 g), HOAc (20 ml) and H_2SO_4 (8 N, 80 ml) contained in a round-bottom flask (250 ml) fitted with a reflux condenser was added dropwise a solution of $KBrO_3$ (8g in 80 ml H_2O) from a separatory funnel while boiling the R.M. After the completion of addition of the oxidant, the reflux was continued for 1 hr. Then the R.M. was cooled and poured onto cold water and extracted with $CHCl_3$ (50 ml × 3). The combined extract (ca 150 ml) was washed successively with H_2O (20 ml × 2) saturated $NaHCO_3$ soln. (10 ml × 2) and H_2O (10 ml × 2), dried (Na_2SO_4), solvent removed and the residue, thus obtained, was weighed and determined the m.p.

(ii) *For substituted benzoic acids* In a round bottom (R.B.) flask (250 ml) were taken the sample (10 g), gl. HOAc (25 ml), and H_2SO_4 (8 N, 100 ml) and the R.M. was heated under reflux while a soln. of $KBrO_3$ (8 g in 100 ml H_2O) was added from a dropping funnel in course of 1.5 hrs. The refluxing was continued for further 0.5 hr. Then the R.M. was cooled and poured onto water and worked up in the usual way with $CHCl_3$ to deliver the desired product which contained the starting material as impurity. Since it was almost impossible to purify the desired acid by fractional crystallisation the following procedure was adopted to obtain the pure product [vide 1.3(e)]. This mixture of impure samples was dissolved in a solution of conc. H_2SO_4 in ethanol (to 100 ml of EtOH was added 5 ml of con, H_2SO_4 and 5 ml of this solution was taken for each gram of the organic acid) and the mixture was heated under reflux for 2 hrs., then cooled and poured onto iced water. The oily ester obtained as the ppt. was extracted with ether (3 times) and the combined organic layer was washed (H_2O), dried (Na_2SO_4) and solvent removed. The residue, thus obtained, was purified and collected through fractional distillation. The high boiling fraction was bromoaromatic ester which on hydrolysis gave the desired bromo-acid in good yield in the following way.

For each gram of the ester was taken 5 ml of aqueous NaOH (4 N), and the R.M. was heated under reflux, then cooled and acidified with conc. HCl. The liberated bromo-acid was collected in the usual way.

Specific examples

- Iodobenzene $\xrightarrow{KBrO_3/HOAc/H_2SO_4 \uparrow\downarrow}$ 4-iodobromobenzene (m.p. 92°C),
- 1,4-Dibromobenzene $\xrightarrow{Do}$ 1,2,4-tribromobenzene (m.p. 87°C),
- Benzoic acid
  (i) $\xrightarrow{Do}$ 3-bromobenzoic acid (m.p. 155°C), (*via* ester).
- Benzanilide $\xrightarrow{\text{usual procedure}}$ 4-bromobenzanilide (m.p. 198°C),
- Nitrobenzene $\xrightarrow{Do}$ 3-bromonitrobenzene (m.p. 56°C),
- Cinnamic acid $\xrightarrow{\text{via ester}}$ 4-bromocinnamic acid (m.p. 253°C),

(b) By Br_2 in CCl_4: (Preparation of dibromoderivative of anthracene). In an Erlenmeyer flask (100 ml) was taken anthracene (0.2 g in 5 ml CCl_4) to which was added excess of a solution of Br_2 in CCl_4 and the R.M. was heated on a steam bath for 15 mins. On cooling the R.M. a ppt. was obtained which was collected in the usual way of filtration under suction, washing, and drying then m.p. noted for identification (m.p. 221°C).

NOTE:

During filtration and washing under suction (usually water aspirator) the solid is collected over the filter paper on the bed of the Buchner funnel. Then the suction is disconnected, the washing solvent is added by a dropper on the solid to just moisten it, again the suction is connected to draw the washing solvent. This process is repeated twice or more to yield the pure solid suitable for future experiments.

(c) By Br_2 in water: (Prepartion of 2,4,6-tribromophenol, m.p. 93°C). In a beaker (250 ml) was taken phenol (4 or 5 drops in 50 ml of H_2O) to which was added an aqueous solution of Br_2 very slowly with constant stirring with a glass rod. When the supernatant solution became yellow permanently (due to excess of Br_2) and the yellowish ppt. settled down, then the addition of Br_2 was stopped. The R.M. was filtered, the solid washed, (Na_2CO_3 soln. followed by H_2O), dried and m.p. noted.

NOTE:

Large beaker was taken because large amount of bromine water was needed for complete reaction. At the end of the reaction, the aqueous layer must assume yellow colour. If excess of Br_2 is not available then a mixture of mono-, di-, and tri-bromophenol would be formed.

(d) By elemental Br_2: (Preparation of dibromocoumarin from coumarin). In an Erlenmeyer flask (25 ml) was taken a soln. of coumarin (0.5 g) in dry ether (5 ml) cooled in ice. One drop of conc. H_2SO_4 is added to it followed by Br_2 (0.5 g) dropwise with shaking. After 5 mins it was taken out of the ice bath and left at room temperature. On evaporation of the ether a sticky residue was obtained which was treated with a little very dilute NaOH soln. when solid dibromocoumarin was obtained. This was filtered, washed, dried and m.p. noted (m.p. 102°).

(e) By Br_2 in glacial HOAc: (Preparation of a *p*-bromoacetanilide, m.p. 167°C). In an Erlenmeyer flask (100 ml) was taken a soln. of acetanilide (1.4 g) in gl. HOAc (4.5 ml) and Br_2 (0.6 ml) in gl. HOAc (2.5 ml) was added dropwise to it with constant shaking. After the completion of addition the R.M. was left at the r.t. for 15–20 mins. Then poured onto cold water (40 ml) when the colourless crystalline ppt. of the desired material was obtained. The pale yellow RM (due to excess of Br_2) was treated with a little $NaHSO_3$ to make it colourless and the ppt. was filtered under suction, washed with H_2O, dried (over steam bath or in an electric oven) and weighed (1.7–1.8 g, ca 85%; m.p. 164–166°C). If required this product may be crystallised from ethanol, m.p. 167°C.

(f) Preparation of iodo-compounds: By I_2 (e.g. methyl iodide from methanol)

(i) In a distillation flask (250 ml; the side tube was closed by a rubber tube using a pinch cock) were taken red P (10 g) and methylalcohol (75 ml) and powdered I_2 (120 g in 10 instalments) was added to it with shaking. After the addition of first portion (12 g) of I_2 the RM was left at r.t. for 3–4 min. Then the next portion (12 g) was added similarly. After the completion of addition of I_2 (120 g), the R.M. was shaken well, stoppered lightly with a cork and left at r.t. for 10–12 hrs.

(ii) The rubber tube closing the side tube of the distillation flask was removed, an air condenser was fitted to the side tube, the free end of the condenser was dipped under iced-water contained in a beaker. On distillation of the R.M. over a steam bath the CH_3I distilled into the receiver under cold water as an oil (pale brown). This oil was transferred to a separatory funnel, washed with dil. NaOH (0.1 N, cold) then H_2O and dried ($CaCl_2$) to furnish almost colourless methyl iodide (110 g, ca 85%). If required it could be distilled (b.p. 43°C, density 2.3, mol. wt. 142) using cold water circulation through the water condenser and a $CaCl_2$ guard tube to maintain the anhydrous condition of the ice-cold receiver.

NOTE: When methanol was replaced by ethanol then ethyl iodide (C_2H_5I, b.p. 72°C, density 1.93, yield ca 80%) was obtained.

(g) By constant boiling HI: (e.g. 2-iodopropane from 2-propanol):

(i) In a distillation flask (250 ml) were taken *iso*-propyl alcohol (15g, 17 ml) and constant boiling HI (225 g, 132 ml; 57%; b.p. 127°C) and the R.M. was slowly distilled using a water condenser (1–2 drops/sec). The distillation was stopped when almost half of the contents of the distillation flask had been collected in the receiver.

(ii) The distillate separated into two layers. The lower layer (impure 2-iodopropane) was washed successively with HCl, H_2O, 5% $NaHCO_3$ and finally H_2O. The purified, almost colourless, oil was dried (anhydrous $CaCl_2$ chips) and distilled to collect pure 2-iodopropane (yield 16 g, 60%, bp. 89°C, density 1.7) under anhydrous condition by fixing a $CaCl_2$ guard tube on the side tube of the receiver. The alkyl iodides affect rubber stopper, so the dry alkyl iodides are stored in brown bottles using glass stopper or cork covered with molten wax (a few pieces of $CaCl_2$ may be used to keep the contents dry).

(h) By I_2 and dil. NaOH: (iodoform from acetone):

(i) In a beaker (500 ml) were taken dil. NaOH (3 g NaOH in 250 ml H_2O) and acetone (1 ml) and to the mixture was added powdered I_2 portion-wise (1 g for 7 times, total I_2 = 7 g) with continuous stirring by a glass rod. When the I_2 from the first instalment was used up completely, the next unstalment was added. After the addition of 7 g of I_2 with stirring was over the R.M. was left a r.t. for 15 mins. when the heavy yellow ppt. of iodoform settled at the bottom of the beaker. Major portion of water was decanted off carefully and the solid residue was filtered under suction, washed with large amount of water and dried to yield iodoform (1.6–1.7 g) having characteristic smell. This fairly pure CHI_3 could be further purified through crystallisation from rectified spirit (m.p. of CHI_3, 119°C).

(i) By KI: (iodobenzene from benzene diazonium salt)

(i) In a distillation flask (500 ml) were taken aniline (9.3 g, 9.3 ml), H_2O (50 ml) and dil. H_2SO_4 (2N, 75 ml) to make a clear solution, then cooled to 5°C in ice.

(ii) In an Erlenmeyer flask (50 ml) was taken $NaNO_2$ (8 g) dissolved in water (25 ml) and cooled in ice (5°C).

(iii) To the aforementioned aniline sulphate soln. cooled in ice was added the freshly prepared cold $NaNO_2$ solution with swirling and occasional testing for excess of HNO_2 in the R.M. by the starch-iodide papers. As soon as the paper indicated excess of HNO_2, the addition of cold $NaNO_2$ was stopped.

(iv) To the benzenediazonium sulphate, thus prepared, was added a soln. of KI (25 g in 35 ml H_2O) and the R.M. was slowly heated when N_2 evolved. After the vigorous evolution of N_2 ended, the R.M. was steam-distilled to funish C_6H_5I as heavy oil collected in a beaker.

(v) This oil was transferred to a separatory funnel, washed with dil. NaOH (0.1 N) and dried over anhydrous $CaCl_2$, then collected by distillation as a colourless heavy oil (16–17 g, ca 80% b.p. 188°C, density 1.83, Mol. wt. 204).

NOTES:

- This is the best method of preparation of aromatic iodocompounds.
- Starch-iodide paper will liberate iodine by HNO_2 (from $NaNO_2 + H^+$). This iodine will turn the starch into characteristic blue colour.

[C] Reduction

(a) By Fe and conc. HCl: (aniline from nitrobenzene)

(i) Iron filing (35 g) was washed with acetone (15 ml) and dried, then immediately transferred into a round bottom flask (500 ml). Nitrobenzene (18 g, 15 ml) was added to it and a reflux condenser was attached to the flask. Then the whole apparatus-set was taken in a fume-chamber (or the apparatus was set up here from the very beginning). A cold water container was kept near.

(ii) Conc. HCl (75 ml, 10 ml each time) was added from the top of the condenser when a vigorous reaction started which should be controlled, if necessary, by cooling the flask in the nearby cold water. After the completion of addition of HCl (75 ml), the flask was heated on a steam bath for 30 mins. when a complex salt composed of aniline HCl, $FeCl_2$ and $FeCl_3$ was formed.

(iii) The R.M., thus obtained, was cooled in ice and made alkaline with conc. NaOH soln. with continuous shaking when aniline was derived.

(iv) This R.M. was steam distilled to separate the aniline as a heavy oil which was collected in the receiver. The oil was transferred to a separatory funnel and extracted with ether (25 ml × 3). The combined organic layer was washed with H_2O (10 ml × 2), dried ($CaCl_2$) and solvent removed. The residue was distilled to furnish pure aniline (10–12 g, b.p. 180–182°C, cf. b.p. nitrobenzene, 210°C).

(b) By Sn and conc. HCl: (*m*-toluidine from *m*-nitrotoluene).

(i) In a hard glass test tube (t.t.) 3 or 4 balls of metallic tin (3 – 4 g) were taken and heated when they melted (m.p. 231°C). This was immediately transferred in a mortar while soft and pressed to make flat pieces of Sn to increase the surface area of Sn.

(ii) In an Erlenmeyer flask (250 ml) was taken *m*-nitrotoluene (4 g) to which was added conc. HCl (30 ml) followed by the chips of Sn prepared before. Immediately a water condenser was attached to the reaction flask. When the initial vigorous reaction was over then the R.M. was heated under reflux till the oily appearance of the substrate disappeared.

(iii) The R.M. was cooled in ice, basified with NaOH (6N) and steam distilled to isolate the desired *m*-toluidine.

(iv) The product may contain a little of *m*-nitrotoluene as impurity. So to remove it the product was dissolved in ether (75 ml) and the organic layer was extracted with cold dil. HCl (10 ml × 2). The acidic extract was cooled in ice and the organic base was regenerated with excess of NaOH soln. This R.M. was again extracted with ether (20 ml × 2) and the ethereal layer was washed with water, dried (Na_2SO_4) and solvent removed in the usual way. The residue, thus obtained, was distilled to furnish pure *m*-toluidine (yield 2.4 g, ca 75%, b.p. 199–201°C; cf b.p. *m*-nitrotoluene, 227°C).

NOTE:

Impure aniline, if produced in (a) above, can be separated from neutral nitrobenzene by the procedure described in (b, iv) above.

(c) By yellow ammonium sulphide $(NH_4)SH$: (*o*-nitroaniline from *o*-dinitrobenzene).

(i) NH_4SH is prepared by passing H_2S through conc. (NH_4) OH and used soon.

(ii) In an Erlenmeyer flask (100 ml) fitted with a water condenser was taken *o*-dinitrobenzene (1 g) to which was added freshly prepared (NH_4) SH (20 ml). The R.M. was heated under reflux for 10 mins, then cooled in ice when the desired product crystallised out. This was filtered and dissolved in hot water and filtered to free it from sulphur impurity. This hot filtrate on cooling in ice yielded *o*-nitroaniline which was collected in the usual way (0.78 g, ca 80%, m.p. 71°C).

(d) By $NaBH_4$, Sodium borohydride (arylmethanol from arylaldehyde)

(i) In an Erlenmeyer flask (250 ml) was taken substituted benzaldehyde (5 g) and pure methanol (40 ml) was added to it. The soln. was cooled in ice and $NaBH_4$ was added portionwise (4 × 0.5 g) with shaking then left at r.t. for 12–16 hrs.

(ii) This reaction mixture was acidified with cold HCl and extracted with a mixture of ether and benzene (1:1, v/v; 25 ml × 3). The combined extract was washed with H_2O (10 ml × 2), dried (Na_2SO_4) and solvent removed on a steam bath. The residue, thus obtained, was distilled to yield the desired aryl methanol (70–90% yield).

NOTE:

$NaBH_4$ is less hazardous than $LiAlH_4$. So it is mostly used during metal hydride reduction.

(e) By $LiAlH_4$, lithium aluminium hydride (very strong reductant).

General procedure: (alcohols from oxo-compounds):

(i) Very carefully $LiAlH_4$ (0.6 g) was added to the anhydrous ether (30 ml) contained in an Erlenmeyer flask (250 ml) fitted with a $CaCl_2$ guard tube. The mixture was shaken well and a soln. of the substrate (0.001 mole) in dry ether (10 ml) was added to it. After the addition of further dry ether (20 ml) the R.M. was left at r.t. for 24-48 hrs.

(ii) The R.M., thus obtained, was treated with water (dropwise) when vigorous gas evolution began and a white ppt. appeared. This was allowed to settle down.

(iii) The upper ethereal layer was separated and the aqueous layer was extracted with ether (10 ml × 6). The combined organic layer was washed with water, dried (Na_2SO_4) and solvent removed to yield the desired product either by crystallisation (if the products were solid) or by distillation (if the products were liquid). Evaporative distillation under reduced pressure was also very effective process to isolate high boiling reduction products. The yields are generally 100%.

NOTE:

It is very dangerous to use $LiAlH_4$ in moist atmosphere. Both $LiAlH_4$ and $NaBH_4$ are used to reduce various functional groups of organic compounds.

(f) By Huang—Minlon Reduction (modified Wolff-Kishner Reduction):

(i) *Alcohols from keto-acetates*: ($[R]^{>CO}_{-O-COCH_3} \rightarrow [R]^{>CH_2}_{-OH}$)

The keto-acetate (0.002 mole), KOH (1.0 g), 95% hydrazine hydrate (2.5 ml) and triethyleneglycol (20 ml) were mixed and heated in an oil bath at 130–135°C for 1 hr. Then the temperature of the oil bath was slowly raised to 190°C and left for 1 hr, and finally at 210-220°C for 3.5 hrs.

The R.M. was cooled and poured onto crushed ice (100 g), then acidified (congo-red indicator) with HCl (6 N) and extracted with CH_2Cl_2 (25 ml × 2). The combined extract was washed with water, dried (Na_2SO_4), solvent removed and the residue (alcohol, desired product) was crystallised from *n*-hexane, if required. The yield of crude desired product was nearly 80%.

(ii) *Hydrocarbons from imino-compounds*: –($[R]{=}NH \rightarrow [R]H_2$).

The imino compound (0.006 mole), KOH (4.22), 95% $H_2N\text{-}NH_2$, H_2O (10 ml) and triethyleneglycol (60 ml) were mixed and heated in an oil both at 130–135° for 1 hr, then the temperature was slowly raised to 210–220°C and left for 3 hrs, at this temperature. The R.M. was cooled and poured onto ice (350 g), acidified with dil. HCl, and extracted with CH_2Cl_2 (75 ml × 4). The combined organic layer was washed with H_2O (25 ml × 4), dried (Na_2SO_4) and solvent removed to furnish the crude product (80–85%) which was crystallised from methanol or other suitable solvents to yield pure product (60–70%).

(iii) *Hydrocarbons from oxo-compounds* ($[R]{>}C = 0 \rightarrow [R] > CH_2$)

By the above-mentioned procedures, [f(i) and f(ii)], various oxo-compounds were reduced to hydrocarbons by the conversion of the oxo-group into methylene group in generally 70-75% yields.

[D] Preparation of Esters

(a) Esters from the reaction of organic acids with alcohols, H_2SO_4 as catalyst.

General procedure

(i) In a dry Erlenmeyer flask (250 ml) was taken anhydrous ethanol (or methanol) (200 ml) and conc. H_2SO_4 (d. 1.84; 10 ml) was added to it dropwise and carefully with swirling, then left at r.t. using a $CaCl_2$ guard tube to protect it from moisture of the atmosphere.

(ii) In a round bottom flask (500 ml) fitted with a $CaCl_2$ guard tube was taken the organic acid (50 g) and the aforementioned acidified alcohol (150 ml; for each gm of the organic acid 3 ml of alcohol) was added to it with shaking to dissolve the organic acid. Then the guard tube was replaced by a water condenser, a few chips of boilizer was added, the $CaCl_2$ guard is fitted on the top of the condenser and R.M. was heated (steam bath) under reflux for 25–30 hrs. Then nearly 20% of the total alcohol was removed by distillation, and the cold R.M. was poured onto large volume of cold water (500–600 ml) when oily and heavy ester was precipitated out.

(iii) The entire mixture was transferred to a separatory funnel (1 L), and extracted with ether (75 ml × 5). The combined extract was washed successively with H_2O (25 ml × 3), saturated $NaHCO_3$ solution (25 ml × 2), again water (25 ml × 3) and finally dried over anhydrous Na_2SO_4.

(iv) On removal of the solvent on a steam bath the desired product was obtained as a colourless oil which was distilled from a distillation flask (125 ml) using a water condenser into the tared receiver fitted with a $CaCl_2$ guard tube. This method is very easily performed particularly with large amount of organic acid, but this is time consuming, hence not followed always. The yield ranges between 80 and 85%.

NOTES:

- The water produced during esterification makes conc. H_2SO_4 gradually diluted. This makes the reaction slow, moreover, the dil. H_2SO_4 hydrolyses the ester formed, thus reducing the yield of the ester. This is the reason why 5%, 10% and 15% alcoholic H_2SO_4 (H_2SO_4: alcohol; v/v) is used to esterify mono-, di-, and tricarboxylic acids, respectively, to prevent dilution of the H_2SO_4 by increasing amount of H_2O liberated.
- The addition of alcoholic H_2SO_4 to the organic acid is better than the addition of H_2SO_4 to the solution of organic acid in alcohol, since in the latter procedure there is a chance of decomposition of the organic acid due to generation of large amount of heat during the addition of conc. H_2SO_4.
- The use of Calcutta Wash Bottle mentioned before will minimise the loss of desired product during transformation from one pot to another.
- The mixture of ether and benzene (1:1, v/v) is used to prevent the loss of expensive ether, but benzene is hazardous to health.

- Sometimes the desired ester may be soluble in water, hence more solvent is used and four extractions will be better than three extractions with organic solvents.
- During qualitative detection of organic acids the sample is taken (0.5 g) in a dry t.t. to which is added dry methanol (1 ml) followed by conc. H_2SO_4 (2 drops) and the R.M. is heated on a steam bath with continuous shaking for 5 mins. Then the cold R.M. is poured onto cold water (10 ml) when the characteristic smell of the organic esters is obtained. In this way, the salicylic acid is identified through preparation of methyl salicylate (oil of wintergreen, smell of iodex, an ointment used to relieve muscular pain) possessing characteristic smell.
- **Specific example:** *Methyl benzoate (Niobe oil) from benzoic acid*
  (i) To the mixture of benzoic acid (12.2 g), anhydrous MeOH (40 ml) conc. H_2SO_4 (2 ml) contained in a round bottom flask fitted with a reflux condenser having a $CaCl_2$ guard tube at the top was added a few boiling chips, and the R.M. was heated under reflux for 8 hrs. on a steam bath.
  (ii) The cold R.M. was poured onto cold H_2O (150 ml) and extracted with a mixture of ether and benzene (1:1, v/v; 25 ml × 4). The combined organic layer was washed successively with H_2O (10 ml × 3), saturated $NaHCO_3$ (20 ml × 2), H_2O (10 ml × 3), dried (Na_2SO_4) and solvent removed. The residue was distilled (yield 10.7 g, ca 80%, bp. 200°C, d, 1.09) to furnish Niobe oil having sweet smell.

NOTES:

- The $NaHCO_3$ extract should always be preserved. If the yield of the ester is poor, then this alkaline extract on acidification (HCl) and extraction will yield the unreacted starting organic acid which may be used again. Moreover, the isolation of unchanged acid would help calculating the "material balance" of the reaction.
- The broken porcelain article is collected and with a pestle mortar this is converted into the size of mastard seed (or lentil) and heated with conc. HNO_3 on a steam bath for 1 hr. The HNO_3 is removed and the small chips are washed with distilled water, dried and preserved in a dry stoppered bottle. This is the boiling chip used 2–3 pieces at an experiment each time before starting the boiling. Heating the boiling chips on Bunsen flame is often recommended. Sometimes small glass balls are used as boilizer but in alkaline solution their utility is doubtful.

(b) By Dean and Stark continuous water separation method: (Diethyl oxalate from oxalic acid, $C_2H_2O_4 \cdot 2H_2O$)

(i) In a round bottom flask (1 L) was taken oxalic acid (250 g) and rectified spirit (85% EtOH, 200 ml) was added to it followed by benzene (250 ml) and a few boilizers. The Dean and Stark apparatus was set at the mouth of the flask (usually B-24 standard joint) and the R.M. was heated under reflux on a

Bunsen burner. The mixture of H_2O, C_6H_6 and EtOH distilled and collected in the receiver of the Dean and Stark apparatus and gradually two separate layers were formed of which the lower layer was of H_2O. The upper organic layer (C_6H_6 and EtOH) returned to the reaction flask. The collected water was removed from time to time and the total quantity was recorded (total calculated volume of H_2O was 140 ml).

(iii) When the water collected was 125 ml, the R.M. was cooled, absolute alcohol (EtOH, 25 ml) was added followed by a few boilizers and again heated. When the H_2O separation stopped the R.M. was cooled, poured onto water (500 ml) and extracted with ether. The organic layer was washed successively with H_2O, dil. NaOH, again H_2O and dried (Na_2SO_4). After removal of the solvent the residue was distilled to collect the desired ester in a receiver fitted with a guard tube (yield 235-245 g, ca 85%, b.p. 185°C).

NOTES:

- Large amount of ester preparation is convenient by this method. Also when the substrate decomposes with conc. H_2SO_4, then this method may be suitable.
- The complete water separation requires 15-16 hrs.
- Any boiling liquid when cooled and to be started again, fresh boilizers must be added each time.
- Before pouring the R.M. into cold water, a portion of benzene (ca 30%) may be removed to make the extraction easier.

(c) By CH_2N_2 (daizomethane): (Preparation of only methyl esters):

Taking an example of an acid having mol. wt. neary 180 and one –COOH group the following method has been described.

(i) In an Erlenmeyer flask (100 ml) aqueous KOH soln. (40%, w/w, 12 ml) was taken, to it distilled ether (40 ml) was added and the mixture cooled in ice (5°C) for 15 mins. Freshly prepared nitrosomethylurea (4 g) was added to it and stirred well with a glass rod when CH_2N_2 was generated imparting a pale yellow colour to the ethereal solution.

(ii) In another Erlenmeyer flask (500 ml) was taken a solution of the organic acid (3.3 g) in distilled ether (50 ml) and the aforementioned CH_2N_2 in ether was added to it very carefully so that the lower layer of KOH solution did not pass into the R.M. There should be a vigorous evolution of N_2 and the R.M. would be colourless. The flask generating CH_2N_2 was repeatedly washed with pure ether (40 ml × 4) and added to the R.M. until the ether layer remained colourless.

(iii) This R.M. was left in the ice bath for nearly an hour, then the excess of CH_2N_2 (if present) was decomposed with few drops of glacial HOAc. The R.M. was transferred to the separatory funnel and washed successively with H_2O (10 ml), saturated $NaHCO_3$ (20 ml × 2), again H_2O (10 ml × 3), then dried (Na_2SO_4). On removal of the solvent in the usual way the residue, thus obtained, was distilled to give the desired methyl ester in atmost 100% yield.

NOTES:

- Diazomethane is a very poisonous gas having a faint sweet smell. It should never be generated or handled outside a good exhaust chamber.
- The method is very quick, work-up is very simple, yield reaches upto 100%, and very suitable for esterification of expensive acids as well as in small quantity. However, N-nitrosomethylurea may explode on long keeping at room temperature.
- The reagents should be proportional to the number of –COOH group in the starting organic acids.
- Commercial ether is stabilized (peroxide free) by addition of different chemicals. So ether is distilled before use particularly when the organic acid is in the range of 0.5 to 1.0 g.
- Before addition the ethereal solution of diazomethane could be transferred to a separatory funnel to make sure that KOH is not passing to the R.M. But there is a chance of loss of CH_2N_2 during the transfer.

(d) By the use of esterification mixture (made from P_2O_5, anhydrous $CuSO_4$ and anhydrous Na_2SO_4 referred to as esterifying agent = EA)

NOTE:

- In this process 1° acids [R–CH_2–COOH or Ar $(CH_2)_n$–CO_2H $n \neq 0$] furnish esters in good yield, while 2° acids (R_2CH-CO_2H) do not give esters in satisfactory yield, and the 3° acids (R_3C–CO_2H) and aromatic acids Ar–COOH do not react at all. So if the substrate contains both 1° and 3° COOH groups, the former will be esterified leaving the 3° (or aromatic) –COOH group untouched.

General procedure:

(i) Ethanol (or methanol) was left on CaO for seven days, then distilled to get dry alcohol.

(ii) $CuSO_4 \cdot 5H_2O$ and $Na_2SO_4 \cdot 10H_2O$ are heated separately to get the colourless $CuSO_4$ (anhydrous) and Na_2SO_4 (anhydrous), respectively.

(iii) In an Erlenmeyer flask (500 ml) fitted with a $CaCl_2$ guard tube were taken P_2O_5 (20 g), $CuSO_4$ (100 g) and Na_2SO_4 (100 g) and mixed well to prepare the E.A.

(iv) In another Erlenmeyer flask (100 ml) fitted with a $CaCl_2$ guard tube was taken the organic acid (0.05 mole) and aforementioned dry ethanol (50 ml) was added to it followed by the E.A. (2.5 g) and left at r.t. for nearly 20 hrs.

(iv) From the R.M. ethanol was removed (30 ml) on a steam bath then the residue was poured onto cold water (100 ml) and the precipitated ester was extracted with ether (25 ml × 3). The combined organic layer was washed successively with H_2O, saturated $NaHCO_3$ solution, again water and dried (Na_2SO_4). On removal of the solvent a residue was obtained which was distilled to collect the desired ester and the yield as well as b.p. recorded properly.

NOTES:

- The $NaHCO_3$ extracted on acidification (HCl) followed by usual work up will deliver the unreacted acid to calculate material balance. In the same way, the half esters (derived form substrates with both 1° and 3° –COOH groups on the same molecule) are isolated and characterised.
- Instead of leaving the R.M. at r.t. for 20 hrs. it may be refluxed on a steam bath for 2 hrs and worked up in the usual way. This method is equally good when the yields are compared, rather this method is better in the sense that it saves much time.
- If the mixture of R–CH_2–COOH and Ar–COOH is given, then they could be easily separated from each other in almost quantitative yield whereas the chromatographic separation may not be so quick and easy. [vide 1.3 (e)]
- *Specific examples:–*

Name of ethyl ester, formula (yield %, b.p. °C), in this order:

(i) Phenyl acetate, Ph–CH_2–CO_2Et (74,221).
(ii) β-Phenyl propionate, Ph–CH_2–CH_2–CO_2Et (67, 238).
(iii) Diethyl succinate, EtO_2 C–$(CH_2)_2$–CO_2Et (61, 212).
(iv) Ethyl formate, EtO_2CH (70, 55).
(v) Diethyl glutarate, EtO_2C–$(CH_2)_3$ – CO_2Et (61, 225).
(vi) Diethyl tartarate EtO_2C–CH(OH)–CH(OH)–CO_2 Et (30, 150/9 mm).
(vii) Ethyl chloroacetate, EtO_2C–CH_2Cl (45, 143).
(viii) Ethyl pivalate, Me_3C–CO_2Et (6, 116).
(ix) Ethyl benzoate, C_6H_5–COOEt (0, –).
(x) Substituted ethyl benzoates (0,–).

[E] Hydrolysis of Esters: (*Carboxylic acids from esters*)

(a) By dilute mineral acids:

(i) In a round bottom flask (250 ml) fitted with a reflux condenser were taken the mixture of ester (0.1 mole) and dil. HCl (20%, 100 ml), heated under reflux for 6–8 hrs, cooled and solid NaCl was added to it. The R.M. was transferred to the separatory funnel and extracted with ether (50 ml × 5). The combined ethereal layer was washed with H_2O, then excess of saturated $NaHCO_3$ (this is preserved, it contains the soluble Na–salt of the organic acid). Care should be exercised so that nothing is lost due to vigorous evolution of CO_2. The ethereal layer was washed (H_2O) free from alkali, dried and the solvent was removed to isolate the unreacted ester, if present.

(ii) The combined $NaHCO_3$ extract (ca 200 ml) was taken in a beaker (large), cooled in ice and acidified with conc. HCl using congo-red as indicator to ensure complete liberation of organic acid which was extracted with ether (50 ml × 5), washed with water (10 ml × 2), dried (Na_2SO_4), solvent removed and the acid isolated to determine the yield and m.p.

NOTES:

- Careless shaking of the $NaHCO_3$ and acid in a separatory funnel may lead to excessive pressure of CO_2 causing ejection of the R.M. like a fountain from the separatory funnel. So from time to time the pressure should be released by opening the stop cock of the separatory funnel.
- During acidification of $NaHCO_3$ extract a beaker would be better than a narrow-mouthed Erlenmeyer flask.
- Generally the solid acids are crystallised form benzene, ether or petroleum ether. Very rarely dil. ethanol, acetone or water is used.
- When large amount of organic acids are prepared which are not steam volatile then the solvent may be removed from a porcelain basin.
- With esters which do not undergo easy hydrolysis with dil. HCl (e.g. cyano-ester and similar compounds), a mixture of glacial HOAc (6 ml for every 1 g ester) and dil. HCl (20%, 6 ml for every 1 g ester) was used. The R.M. was heated under reflux for nearly 40 hrs. The cooled R.M. was extracted with $CHCl_3$ (HOAc is insoluble) and worked-up in the aforementioned way to isolate the desired carboxylic acids.
- It is difficult to isolate the alcohol generated during aqueous inorganic acidic hydrolysis of esters but in alkaline hydrolysis this alcoholic part can be easily collected and identified as shown below.

(b) By dil. KOH (or NaOH) solution (*Alkaline hydrolysis of esters*):

(i) In a round bottom flask (250 ml) fitted with a reflux condenser was taken a mixture of ester (10 g) and dil. NaOH (6 N, 100 ml) and heated under reflux until the R.M. became clear (generally 2 hrs.). Then the R.M. was carefully distilled to collect the distillate which was a mixture of alcohol (from the ester) and water. The residue in the distillation flask was the Na–salt of the organic acid.

(ii) This solid residue was dissolved in minimum volume of water, cooled in ice and acidified (congo-red) with conc.-HCl. The precipitated organic acid, if solid, was filtered, washed free from HCl, dried, weighed and m.p. determined to identify the product. If necessary the acid might be crystallised from benzene or petroleum ether, hot water, dil. EtOH etc. and apposite derivatives may be prepared to confirm the identity of the acid.

(iii) If the organic acid was liquid it was extracted with ether or a mixture of ether and benzene (1:1,v/v) and worked up in the usual way to furnish the desired acid which was identified through b.p. and appropriate derivative(s).

(iv) The alcohol present in the aqueous distillate was identified through the preparation of CHI_3 (directly or after oxidation with red hot copper spiral) in the usual way to identify the alcohol as EtOH or R–CH (OH)–Me (m.p. CHI_3, 119°C). Sensitive colour reactions are also recommended (vide infra).

NOTES:

- If the sample is benzyl benzoate (b.p. 323°C) the R.M. would not be clear or transparent due to the formation of water-insoluble benzyl alcohol.
- For complete hydrolysis of diethyl phthalate more time of refluxing is needed.
- The methanol or ethanol, if obtained after distillation, can also be identified through colour reactions (vide infra).
- Alcohols obtained from ester-hydrolysis can be oxidized and their 2,4-DNP or S-one derivatives will identify the alcohols.
- *Specific examples (Alkaline hydrolysis of methyl salicylate, b.p. 223°C)*
  - (i) In a round bottom flask (RB flask, 100 ml), sample (4 g) was taken and dil. NaOH (4N, 40 ml) was added to it when a white ppt. was formed. The R.M. was heated under reflux for one hour and the clear solution, thus obtained, was cooled.
  - (ii) The R.M. was transferred to a distillation flask and distilled to collect the aqueous solution of MeOH. The residue was Na-salt of salicylic acid which was cooled in ice and acidified with cold (6 N) HCl (congo-red indicator) when colourless ppt. of salicylic acid was formed which was collected (either by filtration or by solvent extraction, the latter method is better), in the usual way and the weight and m.p. determined (usual yield 3.1 g, ca 80%, m.p. 155°C).
  - (iii) The distillate containing methanol was cooled in ice, oxidised with red hot copper wire (spiral) and divided into three parts:

    The part I, furnished a 2,4-DNP (yellow), m.p. 164–166°C.

    The part II gave deep-blue colour with 1% aqueous phenylhydrazine, hydrochloride (0.1 ml) and 0.5% aqueous sodium nitroprusside (0.1 ml) to confirm methanal (derived by oxidation of methanol).

    The third part was treated with 1% phenylhydrzine hydrochloride (4 drops) and aqueous potassium ferricyanide (0.1 ml, 5%) when bright red colour confirmed methanal (derived from methanol).
- *Alkaline hydrolysis of diethyl phthalate, b.p. 296°C*
  - (i) In a RB flask (100 ml) fitted with a reflux condenser was taken ester (2.2 g) and dil. NaOH (6N, 40 ml) was added to it. The R.M. was heated under reflux until a clear solution was obtained, then the R.M. was distilled. The distillate contained ethanol and the residue was sodium phthalate.
  - (ii) The residue was dissolved in cold water and acidified with conc. HCl, avoiding too much dilution of the R.M. The precipitate of phthalic acid was filtered, washed, dried, weighed and m.p. noted, m.p. 193–195°C, (yield 1.1 g, ca 70%).

(iii) The ethanol in the distillate was oxidized to ethanal in the usual way in ice-cold condition by red hot copper spiral and divided into three parts:

To the first part was added I_2 and dil. NaOH to yield CHI_3, m.p. 119°.

To the second part was added 2,4-DNPh-HCl to yield 2,4-DNP. m.p. 148°.

To the third part was added 5% $Na_2Fe(CN)_5NO$ (0.5 ml) followed by piperidine (1 drop) when brilliant blue solution was obtained. If instead of piperidine, one drop of dil. NaOH was added, the rose-red solution would be formed (test for ethanal).

[F] Acetylation

(a) By Ac_2O *(Preparation of β-naphthyl acetate, m.p. 70°C).*

(i) In an Erlenmeyer flask (50 ml) fitted with an air condenser (45 cm) was taken freshly distilled (b.p. 140°C), acetic anhydride (10 g) followed by (a few boiling chips and β-naphthol (5 g) and heated on a low flame to mild boiling (15 mins). The R.M. was cooled and poured onto iced-water (50 ml) and stirred well when β-naphthyl acetate gradually separated which was filtered under suction, washed with cold water dried carefully as the m.p. of the desired product was low (m.p. 70°C) and collected (yield 4–5 g).

NOTES:

- Use of anhydrous NaOAc or $ZnCl_2$ or pyridine (as solvent) or even a drop of conc. H_2SO_4 in the above experiment improves the yield of the desired product.
- The heating should be such that the vapour of Ac_2O rises only upto one-third of the air condenser. M.P. of β-naphthol is 122°C.

(b) By Ac_2O and anhydrous $ZnCl_2$ *(Preparation of mannitol hexaacetate)*

In a RB flask (50 ml) mannitol (5 g), acetic anhydride (20 ml) and anhydrous $ZnCl_2$ (a small amount) were taken, a few boiling chips added to it and the R.M. was heated under reflux for 1 hr. Then the cold R.M. was poured onto cold water and stirred with a glass rod when colourless mannitol hexaacetate was obtained which was filtered under reduced pressure, washed with water, dried and collected (yield 3.5-4 g, ca 50%). This could be further crystallised from glacial HOAc or EtOH (m.p. 119°C).

(c) By Ac_2O and NaOAc on amine hydrochlorides: *(Preparation of acetanilide).*

(i) In an Erlenmeyer flask (100 ml) was taken aniline (1 g) which was dissolved carefully (avoiding excess) in dil. HCl (0.9 ml of HCl in 35 ml of H_2O) when an almost clear solution was obtained.

(ii) To this solution was added fresh Ac_2O (1.5 ml) and immediately added NaOAc (1.2 g). The flask was closed with a rubber stopper, shaken vigorously and the solid desired product, thus formed, was filtered, washed (H_2O), dried (through suction pump), and collected, (m.p. 114°C, yield 1.2 g). This can be recrystallised from hot water or very dilute ethanol.

NOTES:

- This procedure is known as Fieser's method.
- If toluidines are the substrates, excess of Ac_2O and NaOAc should never be used. The desired acetates are soluble in dil. HOAc (from Ac_2O).

(d) By Ac_2O and conc. H_2SO_4: *(Preparation of acetyl salicylic acid (aspirin).*

(i) In a porcelain basin (or a small beaker) mixture of salicylic acid (4.5 g) and fresh Ac_2O (4 ml) was taken and treated with conc. H_2SO_4 (2 drops). The R.M. was heated on a steam bath (ca 80°C, 10 mins) with continuous stirring with a glass rod. The R.M. was removed from the steam bath but the stirring was continued.

(ii) When the R.M. became gradually cold (ca 20 mins) iced water (25 ml) was added to it, stirred well and the solid was immediately filtered on a Buchner funnel under suction. This was washed with cold water, dried under suction over a steam bath, weighed and m.p. noted (yield 4.3 g, ca 75%, m.p. 128–130°C). If required this slightly impure aspirin could be recrystallised from a mixture of HOAc and water (1:1, v/v) to furnish pure product m.p. 135°C.

NOTES:

- The pure product should not give any purple colour with freshly prepared $FeCl_3$ solution.
- The precipitated aspirin should be filtered under suction as soon as possible because the H_2SO_4 present will cause hydrolysis even at r.t.

(e) Preparation of diacetyl derivative of *o*-phenylenediamine:

(i) In a t.t. *o*-phenylenediamine (0.2 g, m.p. 102°C) was mixed with H_2O (10 ml), a drop of HCl and Ac_2O (0.5 ml) was added to it followed immediately by NaOAc (0.5 g). The R.M. was shaken violently and the precipitated acetyl compound was collected in the usual way (m.p. 184–185°C).

(f) Acetate derivative of *p*-nitrophenol: Prepared from sample (0.5 g) and Ac_2O (4 ml) on steam bath (2 hrs.), then poured onto water, worked up, (m.p. 81°C). Or, sample (0.5 g) in Me_2CO mixed with NaOAc in water and Ac_2O, shaken, the solid (m.p. 81°C).

(g) Acetyl derivative of 4-amino-3-nirotoluene: In a t.t. sample (0.3 g), NaOAc (1.5 g) and Ac_2O (15 ml) heated (2 hrs) then poured on water, the solid was isolated and m.p. noted (130°C).

(h) Preparation of acetyl derivative of *o*-nitroaniline: In a dry t.t. the sample (0.2 g, m.p. 72°C) was treated with fresh Ac_2O (1 ml) followed by conc. H_2SO_4 (1 drop) and the R.M. was warmed on a steam bath for 10 mins. Then the cold R.M. was treated with iced water and the precipitated solid was collected in the usual way and dried (m.p. 92°C, crystallised from dil. ethanol).

(i) Preparation of acetyl derivative of *m*-aminobenzoic acid:

(i) In a t.t. the sample (0.2 g, m.p. 174°C), H_2O (5 ml) and dil. HCl (1-2 drops) were taken and fresh Ac_2O (1 ml) was added to the mixture followed by immediate addition of NaOAc (1 g). The R.M. was shaken vigorously and the precipitate, thus obtained, was collected in the usual way. The product was crystallised from dilute ethanol (m.p. 248-250°C).

[G] Preparation of Anhydrides

(a) Preparation of phthalic anhydride by sublimation:

In a small porcelain basin was taken phthalic acid (0.4 g) which was covered with a funnel and the upright stem was plugged with a little cotton to partially block the outlet. The substrate was heated on a sand bath on a low flame when long needle-shaped crystalline phthalic anhydride began to be deposited on the inner-wall of the funnel. Generally, the conversion was complete within 15 mins. (yield 0.3 g, m.p. 133°C).

(b) Preparation of succinic anhydride:

(i) A mixture of succinic acid (5.9 g) and freshly distilled Ac_2O (10 ml) was taken in a round bottom flask (50 ml) fitted with a condenser having a $CaCl_2$ guard tube, and heated on a steam bath (2 hrs) when a clear solution was obtained.

(ii) The R.M. on cooling (ice) deposited crystals of succinic anhydride which on filtration (Buchner funnel), washing with dry ether (4 ml × 2), and drying (vacuum desiccator) furnished the desired product which was stored in sealed bottle (4.5 g, 90%, m.p. 119°C).

[H] Nitration: By conc. HNO_3 and conc. H_2SO_4

Preparation of nitrobenzene from benzene

(i) In a R.B. flask (250 ml) was taken conc. HNO_3 (21 ml). Conc. H_2SO_4 (25 ml) was added to it very carefully and slowly with continuous shaking so that the temperature of the mixture remained at nearly 50°C. If needed the mixture could be cooled in water to control the heat generated.

(ii) Distilled benzene (7.5 g, 8.7 ml) was carefully added dropwise to the mixed acid (prepared in i), shaken well, temperature controlled near 55°C (to prevent the formation of dinitrobenzene) and after 15 mins. interval the second portion of benzene (7.5 g, 8.7 ml) was added similarly, then the R.M. was warmed in a hot water bath at 55-60°C for 40 mins. with occasional shaking.

(iii) The R.M. was poured onto cold water (150 ml) and stirred well with a glass rod when nitrobenzene settled down as a heavy oil. The major portion of the water was decanted off and the pale yellow oil was taken in a separatory funnel, washed with water (20 ml × 3) followed by $NaHCO_3$ solution (saturated, 20 ml × 2) and water (20 ml × 2), dried ($CaCl_2$) and distilled to collect nitrobenzene (yield 15-16 g, ca 70%, b.p. 210°C, d 1.2 g/ml) as a pale yellow oil.

NOTES:

- Nitrobenzene is poisonous and able to penetrate skin.
- Calcutta Wash Bottle should be used during manipulation.
- Two separatory funnels set up side by side are very useful to handle the lower heavy oil (or solvents) when they are required.
- The oil is taken in an Erlenmeyer flask, and anhydrous $CaCl_2$ pieces are added to it, shaken well and the anhydrous oil was transferred to the distillation flask using cotton filtration and Calcutta Wash Bottle with low boiling solvent(s).
- *Preparation of m-dinitrobenzene:* (derivative of nitrobenzene)
  (i) In a hard glass t.t. conc. HNO_3 (2 ml) and conc. H_2SO_4 (2 ml) were mixed well.
  (ii) In a beaker (50 ml) was taken nitrobenzene (1 ml) and the mixed acid (prepared in i) was added to it with stirring by a glass rod. The R.M. was heated carefully on a very small flame (10 mins.) and a drop of R.M. was poured onto cold water (or ice). If solid appeared then that will indicate the completion of nitration and heating should he stopped.
  (iii) The R.M. was poured onto crushed ice contained in a beaker and stirred well when pale yellow solid appeared. This was filtered on a Buchner funnel under water aspirator, washed with water (20 ml × 3) then dried on a blotting paper (or filter paper) to furnish the desired *m*-dinitrobenzene (1.4–1.5 g, ca 85%, m.p. 88°C). If needed it may be crystallised from EtOH (m.p. 90°C).

NOTES:

- Conc. HNO_3 as well as conc. H_2SO_4 are very corrosive chemicals. They should be handled very carefully. Immediate washing with water is required in case of accidental contact with the skin.
- The setting up of water aspirator and its use will be shown by the supervisors.
- During washing of solid material under suction–filtration, the tube is disconnected from the Buchner flask, the desired solvent is added to submerge the solid and the tube is connected again to suck out the washings. This process is repeated again to ensure complete washing to purify the solid material on the Buchner funnel. Incomplete washing will result if the suction is not disconnected.
- If the m.p. of the sample is less than 100°C, it cannot be dried over steam bath (a beaker with boiling water containing the sample on a watch glass covering the mouth of the beaker). So they can be dried under vacuum with controlled temperature (caution, the sample may vapourise and be lost) or better on a blotting paper or filter paper (with small amount of the sample if the sample is explosive).

- *Preparation of p-nitroacetanilide (from acetanilide)*
  (i) In a t.t. conc. HNO_3 (3 ml) and conc. H_2SO_4 (2 ml) were mixed well.
  (ii) In a hard glass t.t. (30 ml) was taken glacial HOAc (3 ml) and acetanilide (2 g) was added to it. The mixture was slowly warmed to dissolve acetanilide, on cooling as soon as the crystallisation started conc. H_2SO_4 (cooled in ice, 4 ml) was added to it and the mixture was shaken well and cooled in ice.
  (iii) The aforementioned mixed acid (in step i) was added dropwise from a dropper to the cold solution of acetanilide (step ii) with stirring by a glass rod keeping the temperature of the R.M. at nearly 5°C (and never above 20°C). After the completion of the addition the R.M. was left at r.t. for 40 mins, then poured onto crushed ice contained in a beaker and stirred well. The precipitated *p*–nitroacetanilide was filtered under suction (Buchner funnel), washed with H_2O (20 ml × 3), dried (steam bath) and collected (2 g, ca 75%, m.p. 210–212°C). It may be crystallised from dil. EtOH.

NOTE:

- If the reaction temperature is not controlled carefully, impure desired product will result due to slow hydrolysis of acetanilide in acidic environment.

(a) By dilute HNO_3: *o-, and p-nitrophenol from phenol*:

(i) In a separatory funnel was taken phenol (25 g) mixed with water (2.5 ml).

(ii) In a R.B. flask (500 ml) was taken a hot solution of $NaNO_3$ (40 g) in H_2O (100 ml), slightly cooled and conc. H_2SO_4 (28 ml) was added to it with shaking. This nitrating mixture was cooled to 20°C.

(iii) The aforementioned phenol (step i) was added dropwise to the nitrating mixture with occasional shaking and left at r.t (20-25°C) for 2 hrs. Then H_2O (200 ml) was added to it and stirred well when a heavy oil separated and settled down. The upper aqueous layer was decanted off very carefully and the residue was washed twice with water to remove the mineral acids from the organic precipitate. Then this is steam distilled when steam volatile *o*-nitrophenol was collected in the receiver as an oil which gradually became solid. This was filtered under suction, washed with water, dried and collected (yield 12-15 g, m.p. 45°C).

(iv) Since *p*-nitrophenol was not steam volatile it stayed in the distillation flask. This material was treated with dil. NaOH (2N, 50 ml) and the light brown solution, thus obtained, was treated with animal charcoal and filtered. The almost colourless filtrate was concentrated by evaporation of the solvent on a steam bath when Na–salt of *p*-nitrophenol crystallised out. This was filtered, washed with dil. NaOH (2N), transferred to a beaker and dissolved in warm water and acidified with dil. HCl when an oil separated which on cooling gradually solidified. This was filtered, washed with H_2O, dried and collected (yield 3–4 g, m.p. 114°C). This *p*-nitrophenol could be crystallised from dilute ethanol.

(b) 2,4,6-Trinitrophenol (Picric acid):

(i) In a porcelain basin (100 ml) was taken phenol (5 g) and conc. H_2SO_4 (13 ml, 24 g) was added to it and warmed in hot water for half an hour when a clear solution was obtained. Water (20 ml) was added to it and the mixture was transferred to a beaker (250 ml) and cooled well (10–15°C).

(ii) To the cold mixture was added conc. HNO_3 (10 ml) stirring all the while with a glass rod when the R.M. became hot gradually and a red colour developed slowly.

(iii) This mixture was left in hot water for 2 hrs. then cooled (10–15°C) when deep yellow picric acid separated. A little cold water was added to the solid and this was filtered under water suction. The solid, thus obtained, was washed with cold H_2O (5 ml × 2) and dried over blotting paper (yield 6 g, m.p. 118–120°). On crystallisation from hot water the pure picric acid was obtained (m.p. 122°C).

NOTE:

- Picric acid when dry is dangerously explosive. It should never be dried with heat. This is stored under water. Just before use it is dried on a blotting paper (or filter paper) with a spatula. It is a mild poison affecting nervous system and very bitter to taste and stains the skin yellow. Unused picric acid is again put under water.

(c) By conc. HNO_3 and conc. H_2SO_4:

- Preparation of 2,4-dinitrochlorobenzene (derivative of chlorobenzene).
  (i) To the hot mixture (steam bath) of chlorobenzene (7 g) and conc. H_2SO_4 (7 ml), contained in a R.B. flask (100 ml), was added with shaking a mixture of conc. HNO_3 (10 ml, dial. 1.4) and conc. H_2SO_4 (7 ml) in such a way that the temperature of the R.M. remained very close to 100°C. After 1 hr, the R.M. was heated to 120°C (sand bath or oil bath) and left at this temperature for 1 hr. more.
  (ii) The R.M. was cooled and poured on crushed ice (150 g) and stirred well with a glass rod when the desired nitrocompound separated as solid material which was filtered under water suction, washed with water and dried on a blotting paper to furnish 2,4-dinitrochlorobenzene (yield 9 g, m.p. 48–52°C). This was crystallized from dilute ethanol (m.p. 53°C).

(d) m-Nitrobenzoic acid *(derivative of benzoic acid)*:

(i) In a porcelain basin conc. H_2SO_4 (10 ml) was taken and warmed to 70°C (steam bath). In a mortar benzoic acid (2 g) was mixed well with KNO_3 (4 g) by grinding the mixture with a pestle. This solid mixture was added in five portions to the hot H_2SO_4 with stirring each time with a glass rod, then the R.M. was left at 85-90°C (steam bath) for 15 mins.

(ii) This R.M. was cooled and poured onto crushed ice (100 g) contained in a beaker and stirred. The solid precipitate, thus obtained, was filtered on a Buchner funnel under suction, washed with cold water and dried to furnish *m*-nitrobenzoic acid (2.5 g, ca 80%, m.p. 133-135°C). This was crystallised from aqueous ethanol to yield pure derivative, m.p. 140°C.

NOTE:

- Generally an acid is crystallised from dilute ethanol (H_2O:EtOH:1:1, v/v) because pure ethanol may produce some ester thereby affecting the quality of the crystalline product.

(e) Trinitroresorcinol *(derivative of resorcinol)*:

(i) In a small beaker resorcinol (1 g) and a mixture of conc. HNO_3 (2 ml) and conc. H_2SO_4 (2 ml) were taken and cooled in ice.

(ii) In a hard glass t.t. another mixture of conc. HNO_3 (2 ml) and conc. H_2SO_4 (2 ml) was taken and cooled in ice then added dropwise (dropper) to the solution of resorcinol (step i) with stirring. After 3 mins. the R.M. was poured on cold water (100 ml) when pale yellow solid separated which was filtered, dried and collected in the usual way (yield ca 1 g). This was crystallised from a mixture of H_2O (20 ml), EtOH (5 ml) and conc. HCl (4 ml) to furnish pure trinitroresorcinol m.p. 175°C.

[I] Picrate: (derivatives of several compounds):

General procedure:

(i) The molecular weight of the sample is calculated and 0.001 mole of the sample is taken.

(ii) The molecular weight of picric acid is 229. So picric acid taken is 0.229 g (carefully weighed, 0.001 mole).

(iii) Carefully weighed sample and picric acid in the molar ratio 1:1 were taken in dry benzene (separate t.t.) and warmed. When they were mixed the colour of the R.M. changed to deep yellow or red. On cooling, crystals of picrate derivative appeared which were filtered, washed with dry benzene and m.p. noted. In some cases the picrate was highly soluble in benzene (as in anthracene), so the warm R.M. was poured in a porcelain basin (100 ml) and benzene was evaporated (preferably from a vacuum desiccator). In such a case, the substrate and the picric acid must be in the molar ratio of 1:1, otherwise the picrate would be impure and m.p. would be low. In some cases the picrate was prepared in absolute ethanol, here also the solutions of the substrate and the picric acid must be saturated.

NOTE:

- The picrates are unstable, with moisture they slowly decompose intc the ingredients, the hydrocarbons are purified through picrates and on chromatography on alumina the pure hydrocarbons are formed.
- *Specific examples*

(a) Picrate of naphthalene (m.w. 128, m.p. 80°C):

(i) In a dry t.t. sample (0.255 g; 0.002 mole) was dissolved in anhydrous EtOH (saturated).

(ii) In another dry t.t. anhydrous picric acid (m.w. 229; 0.460 g, 0.002 mole) was dissolved in anhydrous ethanol (saturated solution).

(iii) These solutions are mixed and heated in a steam bath with shaking (2–3 mins.) and cooled. The yellow crystalline picrate, thus obtained, was filtered on a Buchner funnel, washed with a little cold EtOH (dropper), dried (vacuum) and collected (m.p. 148–149°C). If required it could be crystallised from anhydrous EtOH.

NOTE

- Submit samples of derivatives in a glass tube fitted with a cork marked with Roll No. (students), Sample No., date, names of the derivatives and m.p. of the derivatives. Never submit derivatives in paper pack (particularly on a filter paper) or in glass tube with a cotton plug (or a folded piece of paperplug).

(b) Picrate of anthracene: (m. wt. 178; m.p. 217°C)

(i) In two t.t. sample (0.350 g) in dry benzene (saturated solution) and picric acid (0.450 g) in dry benzene (saturated solution) were made, mixed together, warmed on a steam bath with shaking (2 to 3 mins) and cooled when deep red crystals of the desired picrate were abtained which were filtered, washed with a little dry benzene (dropper), dried and collected (m.p. 136–138°C). If crystals did not appear (excess solvent was taken) the R.M. was poured on a basin and let the solvent evaporate.

NOTE:

- The picrate is very sensitive of moisture which breaks it into the ingredients. So it should be immediately transferred into a sample tube fitted with a cork.

(c) Picrate of phenanthrene: (m. wt. 178, m.p. 100°C).

(i) Sample (0.350 g, 0.002 mole) in anhydrous ethanol (sat. soln. in a dry t.t.).

(ii) Dry picric acid (450 g, 0.002 mole) in anhydrous EtOH (sat. soln. in a dry t.t.).

(iii) These saturated solutions were mixed and heated on a steam bath with shaking for a few minutes, then cooled and the precipitated crystals separated by filtration (water aspirator, Bucnher funnel) and collected m.p. 139-140°C. This was crystallised from anhydrous EtOH to give pure sample, m.p. 143°C.

(d) Picrate of β-naphthylamine (m. wt. 143, m.p. 110°C, hazardous carcinogenic chemical).

(i) Sample (0.28 g) in dry EtOH to make sat. soln. (in a t.t.)

(ii) Dry picric acid (0.45 g) in dry EtOH to make sat. soln (in a dry t.t.)

(iii) These two solutions were mixed, heated (steam bath, 2 mins), cooled and the yellow crystals collected on a Buchner funnel and dried, m.p. 192–194°C.

(e) Picrate of diethylaniline (m. wt. 149, b.p. 215°C).

(i) Sample (0.3 g) in dry EtOH (sat. soln.) in a t.t. (dry).

(ii) Dry picric acid (0.45 g) in anhydrous EtOH (sat. soln. in a dry t.t).

(iii) These solutions were mixed together, heated (steam bath; 3–4 mins), cooled to precipitate deep yellow crystals which were collected in the usual way, m.p. 140°C.

(f) Picrate of β-naphthol (m. wt. 144, m.p. 122°C)

(i) Sample (0.28 g) in absolute EtOH (sat. soln. in dry t.t.).

(ii) Dry picric acid (0.45 g) in dry EtOH (sat. soln. in dry t.t.).

(iii) These solutions were mixed, heated (steam bath, 3–4 mins), cooled and the yellow crystals of the desired product was collected in the usual way, m.p. 154–155°C.

(g) Picrate of carbazole (m. wt. 167, m.p. 246°C).

(i) Sample (0.35 g) in dry EtOH (sat. soln. in a t.t.)

(ii) Dry picric acid (0.45 g) in dry EtOH (sat. soln: in a dry t.t.).

(iii) These solutions are mixed, heated (steam bath, 3–4 mins), cooled and the precipitated crystals are collected in the usual way, m.p. 182°C.

NOTE:

The m.p. of the picrate if upto 5°C less than the reported value is acceptable, but never if it is even 1°C higher than the reported value.

[J] Preparation of Benzoyl Derivatives (by Benzoyl Chloride, m. wt. 140, b.p. 197°C)

(a) Preparation of benzanilide from acetanilide (m.p. 114°C)

(i) In an Erlenmeyer flask (100 ml) was taken acetanilide (1.5 g) and conc. HCl (10 ml) and heated on a small flame after adding a few boilizers. A rubber tube was attached at the top of the condenser and the open end of the rubber tube was placed under the running water from a tap so that the environment was not polluted by HCl fumes.

(ii) After nearly 10 mins the volume of the R.M. was nearly 5 ml when acetanilide underwent hydrolysis to generate aniline hydrochloride.

(iii) This R.M. was cooled, cold water (50 ml) was added to it, neutralised with NaOH (6N, phenolpthalein indicator), then 2 ml of 6 (N) NaOH was added to make the solution weakly alkaline to which was added C_6H_5COCl (1.5 g) dropwise, shaken vigorously and the alkaline character of the R.M. was tested for (red litmus) after each shaking. Then the R.M. was left at r.t. for sometime when solid benzanilide was obtained which was filtered, washed with enough water to make it alkali–free, dried (steam bath) and collected (yield 1.6 g, ca 80%, m.p. 158–160°C). If needed it could be crystallised from dil. EtOH (m.p. 161°C).

NOTES:

- If liquid separates as the product, then it contains excess of benzoyl chloride in which solid benzanilide remains dissolved. In this condition the product is not isolated, but a small amount of NaOH (6N) is again added and shaken well to decompose the benzoyl chloride into sodium benzoate (soluble in water) when the released benzoyl derivative appears as solid.

- Amines (1° and 2°) furnish benzoyl derivatives and phenols yield benzoate derivatives (Ar–O–CO C_6H_5) in this method which is known as Schotten–Baumann method.
- Use of excess of benzoyl chloride should be avoided.

(b) Preparation of benzoyl derivative of aspartic acid [m.p. 278-280°(d)]

(i) In an Erlenmeyer flask (100 ml) was taken sample (1 g) to which was added $NaHCO_3$ (4 g) followed by H_2O (20 ml) when a solution was obtained. Benzoyl chloride (1.5 g) added to this solution and shaken vigorously until the pungent smell of benzoyl chloride was no more perceived. The R.M. was filtered and the filtrate was acidified with cold conc. HCl and the R.M. was cooled in ice when the benzoyl derivative appeared. This was filtered, dried (steam bath) and treated with dry benzene (10 ml), shaken well and again filtered. The residue on the Buchner funnel was washed with dry benzene, dried (steam bath) and m.p. determined (m.p. 180°C). If needed it might be crystallised from hot water.

(c) By the Schotten–Baumann procedure, the following derivatives were prepared:

- *o-Nitroaniline* (m.p. 72°) to benzoyl derivative (m.p. 85°–87°), crystallised from EtOH, m.p. 94°C.
- *p-Nitroaniline* (m.p. 148°C) to benzoyl derivative (m.p. 195°–197°C), crystallised from EtOH; m.p. 199°C.
- *o-Phenylenediamine* (m.p. 102°C) to dibenzoyl derivative (m.p. 290°–295°C), crystallised from ethanol, m.p. 301°C.
- *β-naphthol* (m.p. 122°C) to benzoate (esterification in alkaline medium) derivative (m.p. 107°C).

  In an Erlenmeyer flask (50 ml), a mixture of β naphthol (0.2 g) and NaOH solution (3N, 20 ml) was taken and benzoyl chloride (2–3 drops) was added to it with vigorous shaking. The precipitated benzoate derivative was filtered, washed with enough water, dried in the usual way and collected (m.p. 107°C, after crystallisation from EtOH).
- *p-Nitrophenol* (m.p. 114°C) to benzoate (m.p. 140°C).

In an Erlenmeyer flask (100 ml) was taken a mixture of *p*-nitrophenol (0.5 g), dil. NaOH (6N, 10 ml) and H_2O (10 ml). To this mixture was added benzoyl chloride (3 to 4 drops) and the R.M. was shaken vigorously. The precipitated benzoate derivative was collected in the usual way and crystallised from ethanol to furnish pure benzoate derivative (m.p. 140°C).

NOTES:

- Schotten–Baumann process is actually a method of preparation of esters of phenols in alkaline medium. Hence, these derivative of phenols are

(Ar–O–CO–C_6H_5) benzoate (cf. oate) while those of amines are (Ar–NH–CO–C_6H_5) benzoyl derivatives.

- *m-Nitrotoluene* (b.p. 231°C) to *m*-toluidine (b.p. 203°C) to benzoyl derivation.

  In a hard glass t.t. was taken a few balls of tin (4 g) and heated when it melted (m.p. 231°C). This was taken in a mortar and pressed with a pestle to make flat pieces of tin (to increase surface area).

  In an Erlenmeyer flask (100 ml) a mixture of conc. HCl (10 ml) and *m*-nitrotoluene (1 g) was taken and the pieces of tin were added to it. A reflux condenser was fitted to the flask and the R.M. was heated under reflux until a clear solution of *m*-toluidine hydrochloride was obtained. The R.M. was cooled and basified with dil. NaOH under ice cold condition followed by addition of benzoyl chloride (2 to 3 drops) with vigorous shaking when the bonzoyl derivative was precipitated which was collected and crystallised from EtOH (m.p. 125°C).

- *β-Naphthylamine* (m.p. 110°C) to benzoyl derivative (m.p. 162°C)

  In an Erlenmeyer flask (50 ml) was taken a mixture of sample (0.1 g) and dil. NaOH (6N, 10 ml) to which benzoyl chloride (2 drops) was added and the R.M was vigorously shaken. The precipitated derivative was isolated and crytallised from EtOH (m.p. 162°C).

- *m-Aminobenzoic acid* (m.p. 174°C) to benzoyl derivative (m.p. 248°C)

In an Erlenmeyer flask (50 ml) was taken sample (0.2g) and neutralised with dil. NaOH (1N) followed by addition of benzoyl chloride (2–3 drops) with vigorous shaking. The precipitated derivative was collected and crystallised from EtOH (m.p. 248°C).

NOTE:

- *Phenacyl bromide and phenacyl chloride* (C_6H_5–CO–CH_2Br and C_6H_5–CO–CH_2Cl, respectively).

  These reagents furnish excellent derivatives of carboxylic acids. Hence, they are extensively used to identify the organic acids through the sharp m.p. of the derivatives.

  Preparation of the reagent:

$$\underset{\text{(Acetophenone)}}{C_6H_5{-}CO{-}CH_3} \xrightarrow[0^\circ C]{Br_2 \text{ in ether}} \underset{\text{(Phenacyl bromide)}}{C_6H_5{-}CO{-}CH_2Br}$$

[K] A Few Selected Experiments for the Beginners

(a) Acetyl derivative of diphenylamine (m.p. 53°C)

In a dry t.t. the mixture of the sample (0.5 g), Ac_2O (2 ml) and glacial HOAc (2 ml) was prepared and heated for 15–20 mins on a steam bath with shaking. Then the cold reaction mixture was poured onto cold H_2O (20 ml) and slowly heated on a steam bath (10 mins) until it became clear solution (if not, a small amount of HOAc was again added to make the R.M. clear solution). When the R.M. was cooled in ice, the acetyl derivative appeared which was collected in the usual way and crystallised from EtOH (m.p. 101°C).

(b) Benzanilide (m.p. 163°C) to benzoic acid (m.p. 122°C) and acetanilide via aniline

In an Erlenmeyer flask (50 ml) was taken a mixture of sample (1 g) and 20 ml of dil. H_2SO_4 (70% w/w; to 10 ml of H_2O was added very carefully 10 ml of conc. H_2SO_4) and the mixture was slowly heated (30 mins) on a small flame using an air condenser (30 inches) after adding a few boilizers. On cooling the mixture was poured on ice (10–15 g), stirred well, the precipitate was collected in the usual way and crystallised from water to furnish benzoic acid (m.p. 120–122°C).

The filtrate after the removal of benzoic acid was basified with cold dil. NaOH followed by extraction with ether. The ethereal solution was washed with water and removed on a steam bath. To the residue was added water (10 ml) and just neutralised with dil. HCl (few drops), treated with Ac_2O (4–5 drops) followed by immediate addition of NaOAc (0.5 g) and shaken vigorously (Fieser process) when acetanilide was obtained. This was collected in the usual way and crystallised from dil. EtOH (m.p. 114°C). Alternatively, the aniline derived after the hydrolysis of benzanilide was treated with Ac_2O (1 ml) and conc. H_2SO_4 (1 drop) and the R.M. was heated on a steam bath for 10 mins, then cooled, poured onto ice and the solid acetanilide was collected in the usual way.

NOTES:

- Bezanilide is not readily hydrolysed by boiling conc. HCl. Hence, 70% H_2SO_4 was used which under boiling condition effected the hydrolysis. The progress of the hydrolysis should be tested for occasionally by azo dye test for aniline.
- If benzoic acid does not appear the R.M. should be extracted with $CHCl_3$ which will deliver benzoic acid in the usual way.
- The aniline sulphate in the filtrate (after removal of benzoic acid) can be converted into 2,4,6-tribromoaniline (m.p. 119°C) by treatment with Br_2.

(c) Coumarin (m.p. 68°C) to *o*-methoxy-cis-coumaric acid

In an Erlenmeyer flask (100 ml) was taken coumarin (1.5 g) followed by saturated solution of $Ba(OH)_2$ (25 ml) and the R.M. was heated under reflux (1 hr.) then cooled. To this solution was added Me_2SO_4 (10 ml in 10 instalments, slowly at 5–6 mins. intervals), cooled in ice (30 mins.) and acidified with conc. HCl. The R.M. was extracted with ether (10 ml × 3) and the combined ethereal solution was extracted again with saturated $NaHCO_3$ solution. Thus the coumaric acid derivative was purified through Na–salt, and the aqueous solution of this Na–salt was acidified with conc. HCl (congo-red) to generate *o*-methoxy–*cis*–coumaric acid which was collected in the usual way and crystallised from petroleum ether (m.p. 88°C).

NOTE:

- Me_2SO_4 is a very toxic chemical.

(d) Benzophenone (m.p. 51°C) to benzoic acid

In a hard glass t.t. was taken benzaphenone (1 g) and powdered KOH (2 g) was added to it and the mixture was strongly heated (5 mins). The resulting potassium salt of benzoic acid was dissolved in water, filtered and acidified with conc. HCl (congo-red) when benzoic acid was precipitated. This was collected in the usual way and crystallised from hot water (m.p. 122°C).

(e) Diphenylamine (m.p. 53°C) to N-nitrosodiphenylamine

In a beaker the mixture of sample (1 g), ethanol (5 ml) and conc. HCl (0.8 ml) was warmed when a clear solution was obtained which was cooled to 5°C. To this cold solution was added a solution of $NaNO_2$ (10% aqueous solution, 5 ml) dropwise, then cold H_2O (10 ml) was added to the R.M. and allowed to stand when the pale yellow crystals of the desired product appeared slowly. These crystals were collected in the usual way and crystallised from petroleum ether (60–80°) to furnish pure product (m.p. 67°C).

(f) Ethyl acetoacetate (b.p. 180°C) to its copper complex

In a t.t. the sample (4–5 drops) and the saturated solution of copper acetate in H_2O were taken and shaken vigorously. In 5–6 mins. the pale green crystals of the copper salt began to appear. The product was collected in the usual way (m.p. 190–192°C). Alternatively, in a t.t. copper sulphate pentahydrate (1/4 tea spoonful) was dissolved in H_2O (4 ml) and dil. NH_4OH was added dropwise when a clear deep blue solution was obtained. To this solution was added ethyl-acetoacetate (2–3 drops), shaken well (4–5 mins) when the pale green copper complex separated. This was collected in the usual way (m.p. 191–192°C).

NOTE:

- Addition of excess of NH_4OH to the copper sulphate is avoided.

(g) Quinoxaline derivative from *o*-phenylenediamine (m.p. 102°C) and benzil (m.p. 95°C)

In a dry t.t. benzil (0.2 g) and *o*-phenylenediamine (0.1 g) were taken and heated (steam bath) for 10 mins. When a brown mass was obtained which was crystallised from methanol to yield the desired quinoxaline. This was filtered, washed, dried and m.p. noted (m.p. 125°C).

NOTE:

- Any one of the ingredients can be identified if the other is known.

(h) Sulphanilic acid [m.p. 288°C (d)] to acetanilide and tribromoaniline via aniline

In an Erlenmeyer flask (100 ml) was taken the sample (2 g) and dil. NaOH (4N, 20 ml) was added to it when the sodium salt of the acid was formed. This solution was heated under reflux for 15 mins. Then the aniline formed was collected by distillation. The product was separated into three parts.

One part was carefully neutralised by dil. HCl (avoiding excess) then fresh Ac_2O (0.5 ml) followed by immediate addition of NaOAc (1 g) was effected with vigorous shaking. The precipitated acetanilide was collected and dried in the usual way (m.p. 114°C).

To the second part was added an aqueous solution of Br_2 (a large volume was required, so the reaction was carried out in a large beaker) until the yellow colour of the reagent remained unchanged. The heavy colourless precipitate of 2,4,6-tribromoaniline was obtained which was collected in the usual way (m.p. 119°C). The third part of the distillate was converted into the benzoyl derivative (m.p. 162°C) of aniline by the Schotten–Baumann method.

(i) *m*-Hydroxybenzoic acid (m.p. 200°C) to 2,4,6-tribromophenol and phenyl benzoate via phenol

In a hard glass t.t. the sample (1 g) was mixed with soda lime (3 g) and heated strongly when C_6H_5–ONa was formed. It was dissolved in H_2O and filtered and the clear filtrate was acidified with dil. HCl when phenol was formed. The R.M. was divided into two parts.

The first part was taken in a large beaker and bromine–water was added until the aqueous layer permanently remained yellow and white precipitate of 2,4,6-tribromophenol was formed which was collected in the usual way (m.p. 96°C).

To the second part of the filtrate was added dil. NaOH (1N, 10 ml) followed by benzoyl chloride (1 drop) with vigorous shaking when phenyl benzoate was obtained which was collected in the usual way (m.p. 70°C).

(j) *p*-Bromoaniline (m.p. 66°C) to the benzoyl derivative

In an Erlenmeyer flask (50 ml) was taken the sample (0.2 g) mixed with dil. NaOH (3 N, 10 ml) then benzoyl chloride (2 drops) was added to it and the R.M. was shaken vigorously when the benzoyl derivative of *p*-bromoaniline was obtained (m.p. 204°C).

(k) Derivatives of carbonyl compounds

(i) *m-Nitrobenzaldehyde (m.p. 58°C)*

In an Erlenmeyer flask the mixture of sample (0.5 g) in ethanol (5 ml) and excess of 2,4-dinitrophenylhydrazine hydrochloride in ethanol was taken and heated on a steam bath when the red precipitate of the 2,4-dinitrophenylhydrazone derivative was obtained. This was cooled, filtered, washed with ethanol, dried (steam bath) and collected [m.p. 290–293°C (d)].

NOTES:

- Volatile solvents can be removed on steam bath (boiling water in beaker) but high boiling solvents should be removed in an electric oven.
- Impure 2,4-DNP derivatives (characterised by m.p. which is not sharp and takes a long range of temperature for complete melting) should be purified by dissolving the derivative in dry benzene and passing through a column of neutral Al_2O_3 (30 times of the 2,4-DNP and the length of the column should be at least 7 times of the diameter) when pure 2,4-DNP is obtained in benzene solution.
- *Semicarbazone derivative of m-nitrobenzaldehyde:* To the solution of sample (0.5 g) in EtOH (5 ml) taken in a t.t. excess of semicarbazide hydrochloride and equivalent amount of NaOAc were added and the mixture was heated on a steam bath (10 mins.) On cooling the R.M. colourless crystals of semicarbazone, (S-one), appeared which was collected and dried in the usual way (m.p. 140°C).

NOTE:

- In some cases, the sample was treated with semicarbazide hydrochloride and pyridine (instead of NaOAc) and the mixture was heated on a steam bath (5 mins) then cooled and poured onto cold water to furnish the desired S-one derivative.

(i) Oxidation of *m*-nitrobenzaldehyde

- In an Erlenmeyer flask the mixture of sample (0.2 g), dil. NaOH (6N, 5 ml) and $KMnO_4$ solution was heated carefully (10 mins) then the R.M. was acidified with conc. HCl followed by a pinch of fresh $NaHSO_3$ powder. On cooling the colourless solution crystals of *m*-nitrobenzoic acid appeared which was collected in the usual way (m.p. 141°C).

(ii) Mesityl oxide (b.p. 129°C)

- To the sample (2–3 drops) was added a solution of 2,4-dinitrophenylhydrazine hydrochloride in ethanol and warmed on a steam bath when deep red precipitate of 2,4-DNP derivative was obtained. This was collected in the usual way (m.p. 200°C).
- The sample furnished iodofrom on treatment with I_2 and dil. NaOH in the usual way (m.p. 119°C).

NOTES:

- Mesityl oxide responds to ferric-ferricyanide test as well an Tollen's test (vide infra).
- α,β-Unsaturated carbonyl compounds give deep red 2,4-DNP derivatives while carbonyl compounds which do not have α,β-unsaturation furnish yellow 2,4-DNP in most of the cases.

(iii) Benzil (m.p. 95°C) to 2,4-DNP derivative

- The sample (0.2 g) was taken in a t.t. and dissolved in EtOH (5 ml) and ethanolic solution of 2,4,-dinitrophenylhydrazine hydrochloride was added to it. The R.M. was heated on a steam bath. On cooling orange 2,4,-dinitrophenylhydrozone (i.e. 2,4-DNP) deposited which was collected in the usual way (m.p. 189°C).

(iv) Acetophenone (b.p. 202°C) to its semicarbazone (i.e. S-one).

- The mixture of sample (0.5 g) in methanol (5 ml acetone-free), semicarbazide hydrochloride (0.5 g) and NaOAc (1 g) taken in a t.t. was heated on a steam bath (5 mins). On cooling the R.M. was poured onto cold water and the precipitated S-one was collected (m.p. 198°C).
- In the usual way the sample (0.5 g) was treated with an ethanolic solution of 2,4-dinitrophenylhydrazine sulphate and heated on a steam bath, then cooled when orange 2,4,-DNP was obtained (m.p. 237°C).
- In a beaker (250 ml) were taken the sample (0.2 g) in H_2O (50 ml) and powdered iodine (1.5 g). The mixture was warmed (ca 70°C) and dil. NaOH (1N) was added to it dropwise with stirring. When the I_2 was completely used

up the addition of NaOH was stopped. The heavy iodoform settled down and the water was decanted off. The solid CHI_3 was collected in the usual way and if needed recrystallised from ethanol (m.p. 119°C).

- The sample (0.5 g) was taken in a beaker and dil. NaOH (3N, 10 ml) was added to it followed by dil. $KMnO_4$ solution. The R.M. was heated to boiling, then acidified with conc. HCl, a pinch of powdered $NaHSO_3$ was added to the hot R.M. when it became colourless. On cooling crystals of benzoic acid appeared which was collected in the usual way (m.p. 120–122°C). This could be recrystallised from hot water (m.p. 122°C).

(v) *p*-Chlorobenzaldehyde (m.p. 47°C) to its derivatives

- Sample (0.5 g) in EtOH (5 ml) was mixed with 2,4-dinitrophenylhydrazine hydrochloride in ethanol and heated on a steam bath then cooled. The yellow 2,4-DNP derivative, thus obtained, was crystallised from benzene (m.p. 268–270°C).
- Sample (0.5 g), EtOH (5 ml), semicarbazide hydrochloride (0.5 g) and NaOAc (0.7 g) were mixed and heated on a steam bath. On cooling, the colourless prceipitate of the *S*-one was obtained and crystallised from dil. ethanol (m.p. 229°C).
- Sample (0.5 g) was oxidised with $KMnO_4$ in alkaline medium in the usual way to deliver *p*-chlorobenzoic acid which was collected in the usual way (m.p. 236°C).

(vi) Crotonaldehyde (b.p. 104–105°C) to the derivatives

- In the usual way, the 2,4-DNP was prepared and crystallised from acetone (m.p. 190°C).
- In the usual way, the S-one was prepared and crystallised from hot water (m.p. 198–199°C).

(vii) N,N-Diethylaniline (b.p. 215°C) to nitroso derivative

In a beaker (100 ml) was taken the sample (2 ml) in dil. HCl (10%, 20 ml) and cooled in ice-salt mixture, then a solution of $NaNO_2$ (1.5 g in 2 ml H_2O) was added very slowly (dropwise) to the mixture which was stirred well and left at r.t. for 15 mins. The yellow precipitate obtained was filtered and dissolved in cold water (3 ml) and the solution was cooled in ice. Then it was made alkaline with dil. NaOH when green nitroso derivative of N,N-diaethylaniline was formed which was extracted with ether, and the ethereal layer was washed with H_2O, dried and removed in a current of air. The residue, thus obtained, was crystallised from ethanol to furnish pure nitroso derivative (m.p. 84°C).

(viii) *p*-Bromoacetanilide (m.p. 168°C) to *p*-bromoaniline

In an Erlenmeyer flask (50 ml) the sample (1 g) and conc. HCl (10 ml) were taken and the mixture was heated mildly (5 mins), then cooled in ice and made alkaline with dil. NaOH (6 N) followed by extraction with ether. The ethereal extract was worked up in the usual way and *p*-bromoaniline was obtained (m.p. 66°C).

[L] Derivatives of Carboxylic Acids (Amides and Anilides):

(a) Benzoic acid (m.p. 122°C):

In a small beaker (50 ml) liquor ammonia (3–3.5 ml) was cooled in ice contained in a larger beaker (250 ml) for 15 mins. In a hard glass t.t. a mixture of sample (0.5 g) and $SOCl_2$ (3 ml) was heated on a steam bath when the acid chloride was formed which was divided into two parts.

One part was added to the cooled liquor ammonia and stirred well when the amide was formed. About 10 ml of cold water was added to it and R.M. was filtered and the residue collected in the usual way (benzamide, m.p. 128°C). The other part of the acid chloride was added to the mixture of aniline (3–4 drops) and dil. NaOH (3N, 10 ml) contained in an Erlenmeyer flask (50 ml) and the R.M. was shaken vigorously when the anilide was formed which was collected is the usual way (benzanilide, m.p. 163°C).

If needed the amide and anilide could be crystallised from dil. ethanol.

NOTE:

- Instead of $SOCl_2$ phosphorus pentachloride (PCl_5) could be used in the following way. In a mortar the sample (0.5 g) and PCl_5 (2–3 g) were mixed well with the help of a pestle and the paste (or sometimes oily product), thus obtained, was divided into two parts. One part was treated with coold liqour-ammonia when the amide was obtained and the other part was transferred to the mixture of aniline and dil. NaOH contained in an Erlenmeyer flask and the anilide formed in the usual way was collected. It should be noted that if OH group is present in the molecule of carboxylic acids then PCl_5 cannot be employed for the acid chloride preparation.

(b) *m*-Hydroxybenzoic acid (m.p. 201°C).

The sample (1 g) and $SOCl_2$ (5 ml) reacted in an Erlenmeyer flask to furnish the acid chloride which was converted into *m*-hydroxybenzanilide (m.p. 157°C) crystallised from hot water.

NOTE:

- Sometimes the sample is dissolved in benzene (to ensure better product) and refluxed with $SOCl_2$. The desired product (amide or anilide) stays in benzene which is worked up to furnish the derivatives.

(c) 3,5-Dinitrobenzoic acid (m.p. 204°C)

The mixture of sample (1 g) and $SOCl_2$ (5 ml) was heated under reflux (15 mins.), then cooled and poured onto ice cold liquor ammonia when the amide was formed which was collected in the usual way and crystallised from dilute ethanol (m.p. 183°C).

NOTE:

- The carboxylic acids (with a few exceptions) are soluble in saturated solution of $NaHCO_3$ but the amide and anilide derivatives are always insoluble in $NaHCO_3$ solution. In this way, the identity of the derivative is established even if the m.p. of the derivative is same to that of the sample (i.e. starting material).

(d) Cinnamic acid (m.p. 133°C)

The mixture of sample (0.5 g) and PCl_5 (2 g) was made into paste (mortar-pestle) and ice cold liquor ammonia was added to it and stirred well when the desired amide was formed which was collected in the usual way (m.p. 141°C).

NOTE:

Cinnamic acid furnishes 4-nitrocinnamic acid (m.p. 285°C).

(e) *p*-Chlorobenzoic acid (m.p. 235°C):

The desired acid chloride was prepared either by treating the sample (1 g) with $SOCl_2$ (6 ml) or from sample (1 g) and PCl_5 (2.5 g) in the usual way. This acid chloride was converted into the amide (m.p. 179°C, crystallised from EtOH) and anilide (m.p. 194°C, crystallised from EtOH) in the aforementioned ways.

(f) *m*-Chlorobenzoic acid (m.p. 158°C)

In an Erlenmeyer flask (50 ml) fitted with a reflux condenser was taken the sample (1 g) and $SOCl_2$ (5 ml) was added to it and the R.M. was heated under reflux (b.p. of $SOCl_2$ 78°C) on a steam bath (30 mins.). The acid chloride, thus formed, was converted to amide (m.p. 134°C, crystals from dil. EtOH), and anilide (m.p. 123°C, crystals from dil. EtOH). The anilide was prepared in benzene solution followed by usual work up viz. wash (HCl followed by H_2O), drying (Na_2SO_4) and removal of solvent (steam bath).

[M] Preparation of S-benzylisothiourenium salts (SBT derivatives) of carboxylic acids

General procedure for preparation of SBT derivatives:

(i) In a beaker (50 ml) was prepared a solution (or suspension) of the sample (0.25 g) in H_2O (5 ml).

(ii) A solution of NaOH (1N) was added dropwise to the aqueous acid to bring the pH of the R.M. at close to 7 (phenolphthalein indicator). The appearance of faint pink colour indicated the complete formation of Na–salt of the carboxylic acid. Very dil. HCl soln. (1 or 2 drops) was added to it to discharge the faint pink colour thereby generating a neutral solution of the Na–salt of the carboxylic acid.

(iii) In an Erlenmeyer flask (50 ml) was taken S-benzylisothiourenium salt (1 g) in H_2O (5 ml). This solution was cooled and added to the aforementioned Na–salt of carboxylic acid with shaking and the R.M. was cooled in ice. The precipitate, thus obtained, was filtered (water aspirator), washed (H_2O), dried (method depending on the m.p. of SBT derivatives) and m.p. noted. If the m.p. was lower than that of the reported one then the impure product could be crystallised from water or ethanol to prepare the pure sample.

Thus,

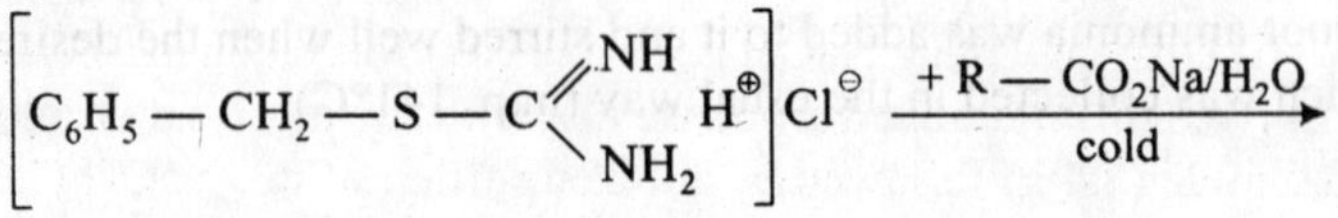

S-Benzylisothiourenium salt

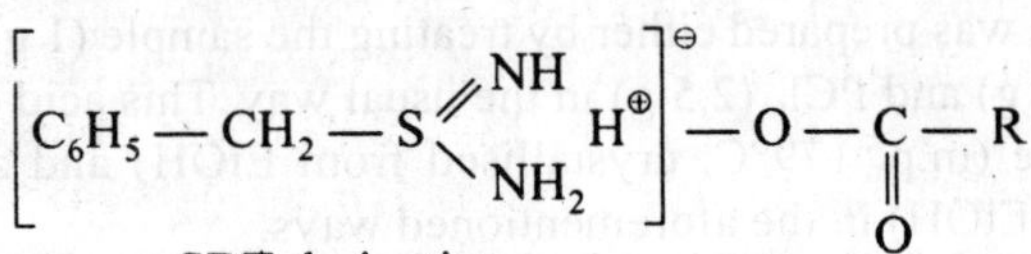

SBT derivative

- *Preparation of the reagent:*

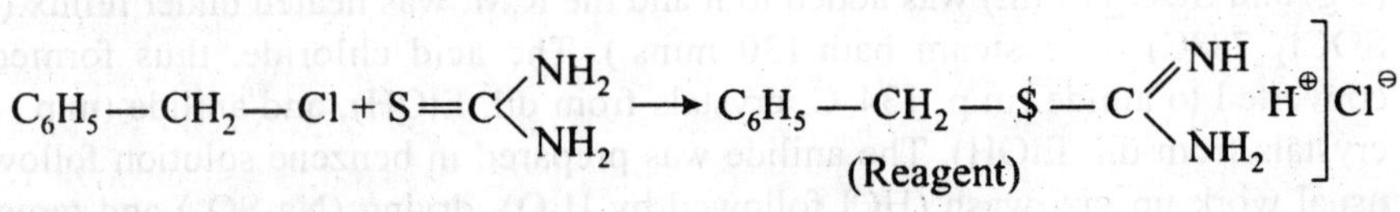

(Benzyl Chloride) (Thiourea)

- *o*-Toluic acid gives SBT derivative (m.p. 146°C).

[N] Preparation of Phthalimide from Phthalic Anhydride (m.p. 133°C)

(i) In a beaker (250 ml) was taken a mixture of phthalic anhydride (15 g) and urea (3.5 g) and heated on a sand bath (180–185°C) when the reaction started with enough frothing. When the frothing stopped a solid product was obtained which was treated with a little cold water and stirred well and filtered under suction, washed (cold H_2O, small quantity) and dried (steam bath). The crude material was crystallised from dil. ethanol (m.p. 238°C, yield of the crude product was 10–12 g).

[O] Preparation of Phthalic Anhydride from Phthalimide via Phthalic Acid

(i) Pthalimide (2 g) was taken in a round-bottom flask (100 ml) fitted with a reflux condenser and dil. NaOH (20%, 20 ml) was added to it. The R.M. was heated under reflux (30 mins.) when the evolution of NH_3 was complete. Then the R.M. was cooled in ice and acidifield with cone. HCl (congo-red). The precipitated phthalic acid was collected in the usual way (m.p. 194–196°C).

(ii) The phthalic acid, thus obtained, was taken in a porcelain basin and covered with a funnel (The stem was partially blocked by cotton) and heated on a sand bath when the long needles of phthalic anhydride deposited on the inner wall of the funnel. The product was collected (m.p. 133°C, yield 1.1 g).

[P] Preparation of Anthranilic Acid (m.p. 145°C) from Phthalimide

Method I. (i) Dil. NaOH (3 g NaOH in 12 ml H_2O) was taken in a beaker and cooled in ice then Br_2 (0.8 ml) was added dropwise to it with stirring. To this cold solution was added phthalimide (2.4 g) followed by another portion of dil. NaOH (2.2 g in 8 ml of H_2O) with stirring. The R.M. was heated at 80°C for 2 mins. then cooled and just neutralised with dil. HCl (congo-red indicator) followed by addition of glacial HOAc (10 ml) then extracted with $CHCl_3$ (20 ml × 3). The combined organic extract was washed with H_2O, dried (Na_2SO_4) and on removal of the solvent on steam bath the crude off white anthranilic acid (0.8 g, m.p. 138–142°C) was obtained which on crystallisation from dil. EtOH furnished pure material (m.p. 144°–145°C).

Method II. In an Erlenmeyer flask (250 ml) was taken dil. NaOH (20 g in 100 ml of H_2O), cooled in ice–salt mixture (–5°C) and Br_2 (5.5 ml) was added to it dropwise followed by a mixture of phthalimide (15 g) and H_2O (15 ml). The R.M. was allowed to attain the room temperature then again NaOH (12g) was added to it and heated at 80°C for a few minutes. The R.M. was filtered and the filtrate was neutralised with conc. HCl (ca 55 ml, congo-red indicator) then treated with glacial HOAc (15 ml). Gradually anthranilic acid was precipitated which was collected through filtration, washing and drying in the usual way. This grey impure product was dissolved in hot water and treated with a little activated charcoal and filtered. This purified solution furnished pure anthranilic acid (10 g, m.p. 145°C).

NOTE:

- Impure anthranilic acid in aqueous solution was treated with Cu $(OAc)_2$ when the copper salt of anthranilic acid was precipitated. This was filtered, washed with water, and the pure solid was suspended in water and H_2S was passed through it. The precipitated CuS was removed by filtration and the clear filtrate was evaporated to dryness (steam bath) to yield pure anthranilic acid.

[Q] Benzaldehyde to Benzyl Alcohol and Benzoic Acid (Cannizzaro Reaction)

(i) Freshly distilled benzaldehyde (10 g) was mixed with a solution of KOH (9 g in 6 ml H_2O) in a flask and shaken vigorously until an emulsion was obtained which was left at room temperature for 15–16 hrs when K–salt of benzoic acid separated.

(ii) The R.M. was diluted with a small amount of H_2O and extracted with ether (25 ml × 3). The aqueous solution was preserved for isolation of benzoic acid. The combined ethereal solution was concentrated by removal of ether and extracted with a solution of $NaHSO_3$ (40% twice extracted) to remove unreacted benzaldehyde. The ethereal solution was washed successively with Na_2CO_3 solution and very little water, then dried (Na_2SO_4). On removal of the solvent a residue was obtained which was distilled to isolate pure benzyl alcohol (3.5–3.8 g, b.p. 205–206°C).

(iii) The aforementioned alkaline solution (in ii) was cooled in ice and acidified with conc. HCl (congo–red indicator) when benzoic acid separated. The R.M. was extracted with ether (15 ml × 3) and the combined ethereal solution was washed with H_2O, and dried (Na_2SO_4). On removal of the solvent crude benzoic acid was obtained (3.5–4.0g) which was crystallised from hot water (m.p. 121–122°C).

NOTE:

- Use of excess of H_2O is to be avoided in this experiment as the work-up of the R.M. becomes very difficult due to presence of benzyl alcohol.

[R] Preparation of Benzyl Alcohol: Cross Cannizzaro Reaction

(i) In an Erlenmeyer flask (250 ml) fitted with a double bored stopper were taken powdered KOH (25 g) and acetone-free MeOH (40 ml) when major portion of KOH was dissolved with generation of heat. A thermometer was inserted through one bore of the stopper, the bulb of the thermometer reached the bottom of the flask, and through the other bore was passed a small reflux condenser.

(ii) The temperature of the R.M. was kept at 60°C (approximately) and a mixture of freshly distilled benzaldehyde (15 g), formalin (15 ml) and acetone-free MeOH (15 ml) was added to it keeping the temperature of the R.M. between 60 and 70°C for 3–4 hrs. Then major portion of MeOH was removed on steam bath.

(iii) To the residual R.M. was added cold H_2O (50 ml) and extracted with benzene (30 ml × 3). The combined benzene layer was washed with H_2O (5 ml × 3), dried (Na_2SO_4) and the solvent was removed. The residue was distilled to isolate pure benzyl alcohol (11 g, ca 75%, b.p. 205–206°C).

[S] Preparation of Quinoline (Skraup's Synthesis)

(i) To the mixture of nitrobenzenc (4 g, 3.3 ml), aniline (6.2 g, 6.2 ml) and anhydrous glycerol (20 g, 17 ml) taken in a round bottom flask (500 ml) was added conc. H_2SO_4 (9 ml) with vigorous shaking, then a reflux condenser was fitted and the R.M. was heated on a wire gauze. As soon as the bubbles began to appear in the hot R.M. it was cooled in cold water to keep the reaction under control whenever required. When the initial vigorous reaction subsided the R.M. was heated under reflux for 3 hrs.

(ii) To the cold R.M. was added H_2O and it was steam distilled to remove nitrobenzene completely (no oily distillate anymore) followed by cooling and addition of excess of NaOH solution very carefully. The R.M. was again steam distilled.

(iii) The aforementioned distillate containing aniline and quinoline was extracted with ether. On removal of the solvent a residue was obtained which was acidified with dil. HCl (conc. HCl, 10 ml in 40 ml of H_2O) and treated with a solution of $ZnCl_2$ [2(N) HCl, 10 ml and 6 g $ZnCl_2$]. On cooling the mixture delivered the double salt of quinoline-zinc chloride which was collected by filtration. The residue was washed carefully with cold HCl (2N), then transferred to a round bottom flask and treated with dil. NaOH to make the R.M. alkaline. On steam distillation quinoline was obtained as oil in the distillate which was extracted with ether. The ethereal layer was washed with H_2O, dried (solid KOH) and solvent removed. The residue was distilled to furnish pure quinoline (4.5–5 g, b.p. 237°C).

NOTES:

- Glycerol was taken in a porcelain basin and very carefully heated directly over a small flame in a fume chamber. When the temperature of the liquid reached 180°C then the glycerine was considered as anhydrous. Stirring the liquid with a glass rod is recommended.
- Use of half portion of the required H_2SO_4 at the beginning of the experiment would ensure better control of the heat of reaction. The other half of the H_2SO_4 could be added when the reaction became mild.

[T] Preparation of Phenol from Aniline (via Diazonium Salt)

(i) In an Erlenmeyer flask (250 ml) were taken conc. HCl (12N, 30 ml), aniline (10 ml) and cooled in ice–salt mixture to 10°C. A solution of $NaNO_2$ (8 g in 20 ml of H_2O) was added to it portion-wise with stirring taking care that $NaNO_2$ was never in excess (tested with KI/starch paper). As soon as the starch paper became deep blue (indicating HNO_2 in excess) the addition of $NaNO_2$ was stopped. In this way, benzene diazonium chloride was prepared.

(ii) The R.M., thus obtained, was warmed at 55°C (warm water) when N_2 gas began to evolve which stopped soon. Then the R.M. was steam distilled to furnish phenol in the distillate which was extracted with ether (30 ml × 3). The combined ethereal solution was washed with small amount of H_2O, and dried (Na_2SO_4) and solvent removed. The residue was distilled to yield phenol (6–7 g, b.p. 182°–183°C).

NOTE:

- Phenol is a corrosive chemical. Care should be taken to handle it

[U] Preparation of Chlorobenzene from Aniline (Sandmeyer Reaction)

(i) Benzene diazonium chloride was prepared from aniline (10 ml) in the aforementioned method (vide T, i).

(ii) In an Erlenmeyer flask were taken $CuSO_4 \cdot 5H_2O$ (15 g), NaCl (5 g) and H_2O (75 ml) and $NaHSO_3$ (8 g) was added to it with vigorous shaking followed by cooling in ice when precipitate of Cu_2Cl_2 was formed which was collected by filtration and dissolved in conc. HCl (30 ml) in an Erlenmeyer flask fitted with a reflux condenser and the solution was heated to 80°C.

(iii) To this Cu_2Cl_2 solution was added the benzenediazonium chloride dropwise with shaking and the R.M. was heated at 80°C for 15 mins. The chlorobenzene, thus formed, was steam distilled followed by work-up with ether to yield the desired product through distillation (8–9 g, b.p. 130°–132°C).

[V] Preparation of p,p′-dichlorodiphenyltrichloroethane (DDT)

In a hard glass bottle with a stopper was taken a mixture of chloral hydrate (4.2 g) and chlorobenzene (5 ml) and warmed a little to make a clear solution which was treated with conc. H_2SO_4 (45 ml) very carefully. This R.M. was vigorously shaken for 2 hrs. at room temperature. Then the R.M. was poured onto H_2O (200 ml) and the precipitates, thus formed, was filtered (water aspirator), washed successively with water, solution of Na_2CO_3, again water, then dried to collect the desired product which was not pure, but generally contained nearly 90% of the *p*-isomer and the m.p. was above 95°C.

[W] Preparation of Salicylaldehyde from Phenol and Chloroform (Reimer-Tiemann Reaction)

(i) To the solution of NaOH (40 g in 40 ml of H_2O) contained in a round bottom flask (500 ml) fitted with a double-bored cork was added freshly distilled phenol (13 g) in hot condition and taking care that sodium phenoxide did not crystallise out at 60–65°C during cooling. A condenser and a thermometer were passed through the stopper and the bulb of the thermometer reached the bottom of the flask.

(ii) Through the top of the condenser was added $CHCl_3$ (10 g, 6.6 ml) to the hot solution of phenol with shaking and after 10 mins. at 60–70°C, the R.M. was again treated with $CHCl_3$ (10 g) with shaking for 15 mins. at 60–70°C and finally the last portion of $CHCl_3$ (10 g) was added to the R.M. in the similar way then heated under reflux for 1 hr. on a steam bath. Excess of $CHCl_3$ was removed by steam distillation.

(iii) The dark residue, thus obtained, was cooled and slowly acidified with dil. H_2SO_4 when it became almost colourless. This mixture was steam distilled as before to isolate phenol and salicylaldehyde in the distillate which was immediately extracted with ether (20 ml × 3). The combined ethereal layer was washed with H_2O and solvent removed on a steam bath. The residue was transferred to a hard glass bottle fitted with a glass stopper, treated with very concentrated solution of $NaHSO_3$ (excess) and the R.M. was vigorously shaken for 1 hr. The precipitated bisulphite salt of salicylaldehyde was filtered, washed with a little alcohol and ether (to remove phenol, if present), dried and transferred to an Erlenmeyer flask then acidified with dil. H_2SO_4.

(iv) The generated salicylaldehyde was extracted in the usual way with ether, washed with water, dried (Na_2SO_4). On removal of the solvent the residue obtained was distilled to furnish the desired salicylaldehyde (5–6 g, b.p. 196°C).

NOTE:

- If in the step (i) of the reaction the sodium phenoxide (C_6H_5–ONa) crystallises out then no Reimer-Tiemann reaction will follow.

[X] Preparation of Cinnamic Acid (Perkin Reaction)

(i) A mixture of freshly distilled benzaldehyde (10.5 g), acctic anhydride (15 g) and dry NaOAc (5 g) was taken in a round bottom flask (50 ml) fitted with an air condenser (ca 30" long) and heated in an oil bath at 180°C for 8 hrs.

(ii) The R.M. was diluted with a small amount of H_2O and transferred to a larger flask followed by steam distillation to remove the residual benzaldehyde, if present. The impure cinnamic acid remaining in the distillation flask was dissolved, in hot water and treated with a little charcoal and filtered. The colourless filtrate was cooled in ice when crystals of cinnamic and was obtained which were filtered, washed with water, dried and collected (ca 6.5 g, mp. 133°C). If needed it may be recrystallised from hot water.

NOTES:

- In a porcelain basin was taken NaOAc · $3H_2O$ and heated directly on a flame when the water of crystallisation evaporated and a solid mass was formed. On continued heating the solid melted which was immediately transferred to a mortar and ground well with a pestle to obtain powdered anhydrous NaOAc which was at once stored in a bottle with rubber stopper. It is advisable to prepare the anhydrous NaOAc just before use.
- When the R.M. is transferred to a larger flask (250 ml) it is better to use a distillation flask (250 ml) to make the steam distillation more easy.

[Y] Preparation of the Adduct from Anthracene and Maleic Anhydride (Diels–Alder Reaction)

In a round bottom flask (50 ml) was taken a mixture of xylene (12 ml), anthracene (1 g) and maleic anhydride (0.55 g) and heated under reflux for 15 mins. On cooling the R.M to room temperature the desired "adduct" was deposited which was filtered and crystallised from xylene to purify the "adduct" then dried at 100°C (steam bath or electric oven), yield 1.5 g, m.p. 255–256°C.

NOTE:

- Sometimes this adduct is referred to as "butterfly compound" based on its structural formula.

[Z] Preparation of Suphanilic Acid

(i) To the mixture of aniline (2 ml) and conc. H_2SO_4 (4 ml) contained in a hard glass t.t. was added very carefully and slowly fuming H_2SO_4 (10% Oleum, 4 ml). The R.M. was shaken well and heated at 180°C for 30 mins. in a sand bath.

(iii) Then the R.M. was poured onto cold water and stirred well when the crystals of sulphanilic acid appeared slowly. This was filtered and dried to yield the desired compound [2.8 g, m.p. near 288°C (d)]. If required, it may be recrystallised from hot water to furnish pure sulphanilic acid dihydrate.

[AA] Ethyl Acetoacetate to 7-hydroxy—4—methylcoumarin

(i) In a t.t. was taken resorcinol (0.45 g) and cooled in ice. Elthyl acetocetate (0.45g) was added to it and shaken well.

(ii) In a separate t.t. conc. H_2SO_4 (1.5 ml) was taken and cooled in ice.

(iii) This cooled H_2SO_4 was added to the cold mixture of the substrate and stirred with a glass rod for half an hour then poured onto crushed ice when the precipitate of the desired product was formed. This product was filtered, washed (H_2O) and crystallised from ethanol to furnish pure material (m.p. 188°–190°C).

NOTE:

- This coumarin derivative may be used to identify either resorcinol or ethyl acetoacetate as unknown compound.

[AB] Preparation of Anthrone from Anthraquinone

(i) In an Erlenmeyer flask (50 ml) fitted with a reflux condenser was taken anthraquinone (1g) and Sn (1g) was added to it followed by glacial HOAc (8 ml). The R.M. was heated under reflux for 2 hrs. while adding conc. HCl (3 ml) from time to time in small portions.

(ii) The R.M. was filtered and the filtrate was cooled when a solid material appeared. This was filtered and washed with water to furnish anthrone which was crystallised from ethanol (yield 0.72 g, mp. 154–155°C).

NOTE:

- Anthrone may be submitted as a derivative of anthraquinone.

[AC] Preparation of Benzoin from Benzaldehyde (Benzoin Condensation)

A mixture of freshly distilled benzaldehyde (7.5 ml), aqueous solution of KCN (1.9 g KCN in 5 ml of H_2O) and rectified spirit (20 ml) was taker in a round bottom flask (100 ml) fitted with a reflux condenser. A few boilizers were added to the R.M. which was heated under reflux on a steam bath for 45 mins. Cold water (75 ml) was added to this cold R.M. with vigorous shaking when the yellow precipitate of benzoin was obtained which was filtered on a Buchner funnel (water aspirator), washed with H_2O, dried (6 g, ca 80%) and crystallised from rectified spirit (m.p. 137°C).

NOTES:

- KCN is a dangerous poison, it should be handled under the supervision of the expert person.

- While adding H_2O to the R.M., it should be continuously shaken, otherwise a big lump of the desired product will be formed which will pose severe problem of taking it out of the reaction flask.

[AD] Preparation of Benzil (Oxidation of Benzoin by HNO_3):

In a round bottom flask (100 ml) was taken a mixture of benzoin (6 g) and conc. HNO_3 (10 ml, d 1.4) and heated on a steam bath for 2 hrs. with occasional shaking. Then the R.M. was cooled and poured onto H_2O (50 ml) and the precipitated yellow benzil was filtered on a Buchner funnel under suction, washed with water, dried and crystallised from ethanol to furnish the desired pure product (5–5.5 g, m.p. 95°C).

NOTE:

- The reaction should be carried out in a fume chamber to avoid environment pollution.

[AE] Preparation of Benzilic Acid (Benzil–Benzilic Acid Rearrangement):

(i) A mixture of benzil (4 g), rectified spirit (10 ml) and aqueous KOH (4 g KOH in 8 ml H_2O) was taken in a round bottom flask (50 ml) and heated on a steam bath for 10–15 mins., then transferred to a porcelain basin and left at room temperature for 15–16 hrs.

(ii) The K–salt of benzilic acid, thus obtained, was filtered under suction, washed with anhydrous ethanol and taken in a beaker. This purified K–salt was dissolved in small amount of water, cooled and acidified (congo-red) with conc. HCl when impure benzilic acid was formed. This was filtered and the solid was dissolved in hot water and treated with animal charcoal and filtered. The almost colourless filtrate was cooled in ice when pure benzilic acid crystallised out which was collected in the usual way (3 g. m.p. 150°C). If needed benzilic acid could be recrystallised from benzene.

NOTE:

- In the same procedure 9-carboxy-9-hydroxyfluorene can be prepared from 9,10-phenanthrenequinone.

[AF] Preparation of 4-oxo-4-phenylbutanoic Acid (Friedel–Crafts Reaction):

(i) Under the exhaust chamber a mixture of Na–dried C_6H_6 (22 ml) and succinic anhydride (3.4 g) was taken in a R.B. flask (100 ml) fitted with a condenser having a $CaCl_2$ guard tube.

(ii) Powdered anhydrous $AlCl_3$ (10 g) was added to the above mixture in one portion when the reaction started immediately. The R.M. was heated under reflux for 30 mins., then it was cooled and treated with H_2O (dropwise, 15 ml) followed by conc. HCl (5 ml), then steam distilled to remove completely excess of benzene.

(iii) On cooling the residue, thus obtained, the desired ketoacid solidified which was filtered, washed with dil. HCl (3N, 10 ml), then H_2O (5 ml × 2) and finally dissolved in aqueous solution of Na_2CO_3 (4 g in 25 ml of H_2O) by heating and filtered while hot.

(iv) The hot filtrate was treated with a little animal charcoal, filtered, and carefully acidified in a large beaker (250 ml) with conc. HCl (7 ml).

(v) The R.M. was cooled in ice–salt bath and the precipitated ketoacid was filtered (Buchner funnel), washed (H_2O), then dried on a blotting paper (or in a vacuum desiccator), then stored (5.5 g, m.p. 115°C).

[AG] Preparation of 4-phenylbutanoic Acid (Clemmensen Reduction):

(i) In a R.B. flask (100 ml) were taken zinc wool (10 g), $HgCl_2$ (1.8 g, *Poison*), conc. HCl (0.5 ml) and H_2O (10 ml) and stirred well for 5 mins. Then the aqueous solution was decanted and the amalgamated zinc, thus obtained, was treated immediately with H_2O (6 ml), conc. HCl (15 ml), pure toluene (10 ml) in this order, followed by 4-oxo-4-phenylbutanoic acid (4.2 g).

(ii) The R.M. was heated under reflux for 30 hrs. under an exhaust chamber adding 5 ml. of conc. HCl at the interval of every six hours.

(iii) The R.M. was then cooled, treated with H_2O (20 ml) and extracted with ether (3 × 10 ml). The combined organic extract was washed (H_2O), dried (Na_2SO_4) and solvent removed on a steam bath.

(iv) The residue, thus obtained, was transferred to a Claisen flask and distilled under reduced pressure using a capillary tube to isolate the desired product (3.2 g, b.p. 178–182°C/20 mm, m.p. 48°C).

[AH] Preparation of α-tetralone (Friedel Crafts Cyclisation):

(i) In a R.B. flask (50 ml) was taken a mixture of 4-pheylbutanoic acid (3 g) and $SOCl_2$ (2 ml) and heated on a steam bath (30 mins) then the excess of $SOCl_2$ was removed carefully under suction. In this way, the intermediate acid chloride was prepared.

(ii) This acid chloride was dissolved in pure CS_2 (20 ml) and cooled in ice then treated with powdered anhydrous $AlCl_3$ (3 g) and left at room temperature till the evolution of HCl gas was complete.

(iii) The R.M. was heated on a steam bath (15 mins.). Then cooled (ice), treated with crushed ice (10 g) followed by conc. HCl (2.5 ml) and finally steam distilled. The first distillate was CS_2 which was rejected.

(iv) The second distillate was extracted with pure benzene (10 ml × 3). The combined organic layer was dried (Na_2SO_4) and solvent removed. The residue was distilled to collect α-tetralone (2 g, b.p. 134°–137°C/15 mm).

NOTE:

- **Distillation under reduced pressure (water aspirator or vacuum pump) cannot be carried out using boilizer. Here a fine capillary drawn from a capillary tube should be used to ensure smooth distillation. This fine capillary should reach the bottom of the distillation flask (usually Claisen flask with a Vigreux column having a mercury well on top for the thermometer).**

[1.7] MINOR PROJECTS FOR THE BEGINNERS (STEPWISE VARIOUS TYPES OF REACTIONS):

The starting material of each project should not be more than 10 g.

(a) Benzaldehyde $\xrightarrow{1}$ Benzoic acid $\xrightarrow{2}$ Benzamide $\xrightarrow{3}$ Aniline $\xrightarrow{4}$ Acetanilide.

(b) Benzaldehyde $\xrightarrow{5}$ Benzoin $\xrightarrow{6}$ Benzil $\xrightarrow{7}$ Benzilic acid $\xrightarrow{8}$ Benzophenone

(c) Phenanthrene $\xrightarrow{9}$ 9,10,-Phenanthrenequinone $\xrightarrow{10}$ 9-carboxy-9-hydroxyfluorene $\xrightarrow{11}$ Fluorenone $\xrightarrow{12}$ Fluorene

(d) Benzene $\xrightarrow{13}$ Nitrobenzene $\xrightarrow{14}$ Aniline $\xrightarrow{15}$ Chlorobenzene $\xrightarrow{16}$ D.D.T.

(e) Succinic acid $\xrightarrow{17}$ Succinic anhydride $\xrightarrow{18}$ 4-oxo-4-phenylbutanoic $\xrightarrow{19}$ 4-Phenylbutanoic acid $\xrightarrow{20}$ α-Tetralone

(f) Aniline $\xrightarrow{21}$ Acetanilide $\xrightarrow{22}$ *p*-Bromoacetanilide $\xrightarrow{23}$ *p*-Bromoaniline $\xrightarrow{24}$ *p*–Bromobenzanilide.

(g) Aniline $\xrightarrow{25}$ Phenol $\xrightarrow{26}$ Salicylaldehyde $\xrightarrow{27}$ Salicylic acid $\xrightarrow{28}$ oil of Wintergreen.

(h) Phthalic acid $\xrightarrow{29}$ Phthalic anhydride $\xrightarrow{30}$ Phthalimide $\xrightarrow{31}$ Anthranilic acid $\xrightarrow{32}$ Ethyl anthranilate.

(i) Benzaldehyde $\xrightarrow{33}$ Benzyl alcohol $\xrightarrow{34}$ Benzyl chloride $\xrightarrow{35}$ Benzyl cyanide $\xrightarrow{36}$ 1-Oxo-2-phenylethane.

(j) Niobe oil $\xrightarrow{37}$ Benzoic acid $\xrightarrow{38}$ Benzamide $\xrightarrow{39}$ Aniline $\xrightarrow{40}$ Sulphanilic acid $\xrightarrow{41}$ Methyl orange

Legends:

(1) Oxidation with $KMnO_4/HO^{\ominus}$

(2) Amide through Benzoyl Chloride.

(3) Hofmann Degradation.
(4) Acetylation (Fieser).
(5) Benzoin Condensation.
(6) Oxidation with HNO_3.
(7) Benzil–Benzilic acid Rearrangement.
(8) Oxidation with $KBrO_3$/HOAc.
(9) Oxidation with $KBrO_3$/HOAc.
(10) Benzil–Benzilic acid Rearrangement.
(11) Oxidation with $KBrO_3$/HOAc.
(12) Huang–Minlon Reduction.
(13) Nitration.
(14) Reduction with Sn/HCl.
(15) Sandmeyer Reaction.
(16) Condensation (water elimination).
(17) Anhydride formation.
(18) Friedel–Crafts Reaction.
(19) Clemmensen Reduction
(20) Cyclisation.
(21) Acetylation.
(22) Bromination.
(23) Hydrolysis.
(24) Benzoylation.
(25) Diazotisation followed by decomposition.
(26) Reimer–Tiemann Reaction.
(27) Oxidation.
(28) Esterification.
(29) Anhydride.
(30) Substitution.
(31) Hofmann Degradation.
(32) Esterification.
(33) Cross Cannizzaro Reaction.
(34) Chlorine Substitution.
(35) Substitution.
(36) Stephen Reduction.
(37) Hydrolysis.
(38) Amide Formation.
(39) Hofmann Degradation.
(40) Sulphonation.
(41) Azo-coupling.

[1.8] TRANSFORMATIONS OF BENZENE DERIVATIVES

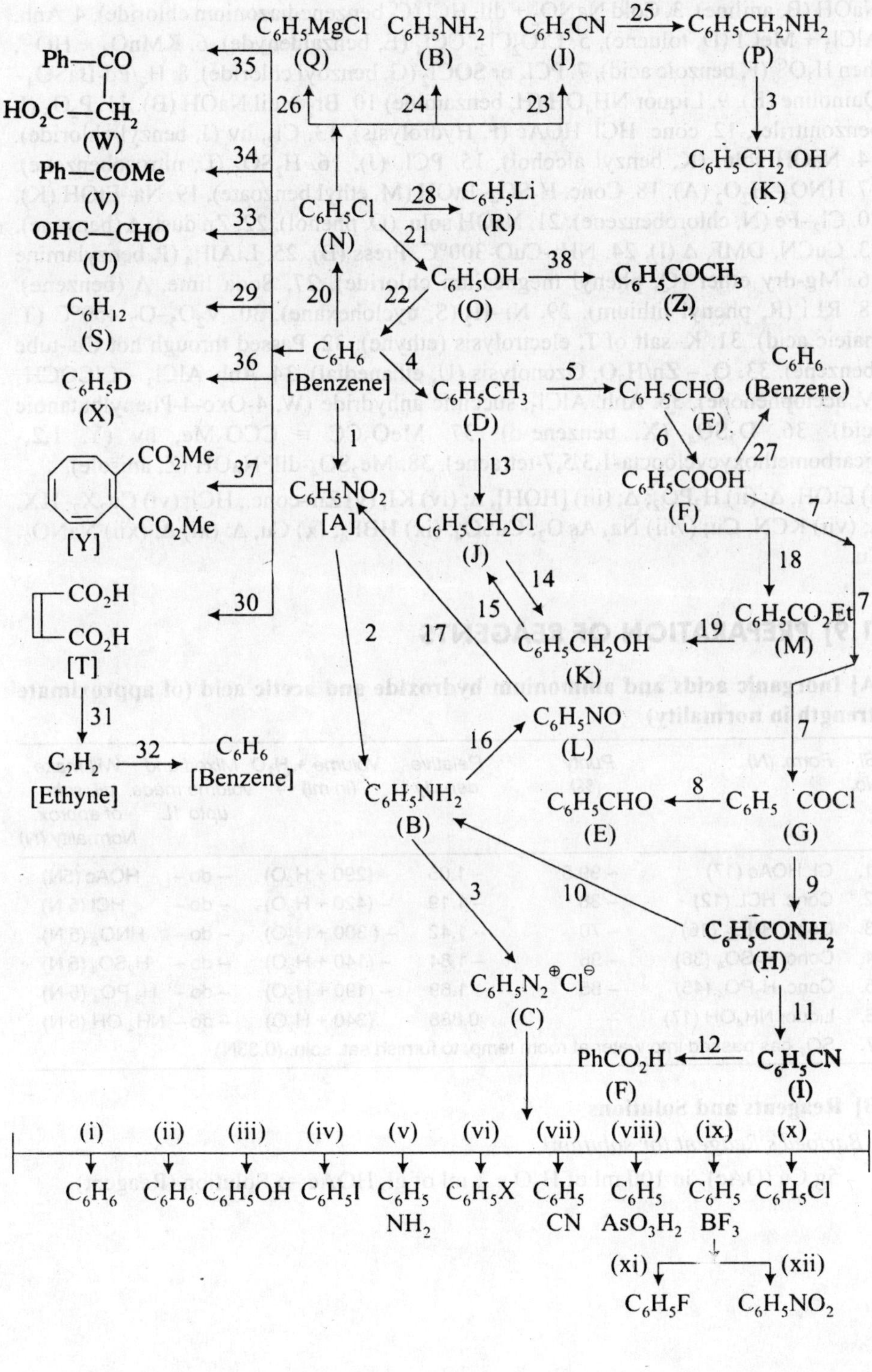

Legends: 1. Conc. HNO_3 + Conc. H_2SO_4 (A, nitrobenzene), 2. Sn + conc. HCl, then NaOH (B, aniline), 3. Cold $NaNO_2$ + dil. HCl (C, benzenediazonium chloride), 4. Anh. $AlCl_3$ + MeCl (D, toluene), 5. CrO_2Cl_2, CCl_4 (E, benzaldehyde), 6. $KMnO_4$ + $HO^{\ominus}$, then $H_3O^{\oplus}$ (F, benzoic acid), 7. PCl_5 or $SOCl_2$ (G, benzoyl chloride), 8. H_2/Pd-$BaSO_4$–Quinoline (E), 9. Liquor NH_4OH (H, benzamide) 10. Br_2 + dil.NaOH (B), 11. P_2O_5 (I, benzonitrile), 12. conc. HCl–HOAc (F, Hydrolysis), 13. Cl_2, hν (J, benzyl chloride). 14. NaOH soln. (K, benzyl alcohol). 15. PCl_5 (J), 16. H_2SO_5 (L, nitrosobenzene). 17. HNO_3–H_2O_2 (A). 18. Conc. H_2SO_4–EtOH (M, ethyl benzoate), 19. Na–EtOH (K). 20. Cl_2–Fe (N, chlorobenzene). 21. NaOH soln. (O, phenol). 22. Zn dust, Δ (benzene). 23. CuCN, DMF, Δ (I). 24. NH_3-CuO-300°C, Press (B). 25. $LiAlH_4$ (P, benzylamine 26. Mg-dry ether (Q, phenyl megnesium chloride). 27. Soda lime, Δ (benzene). 28. RLi (R, phenyl lithium). 29. Ni–H_2 (S, cyclohexane). 30. V_2O_5–O_2 450°C (T, maleic acid). 31. K–salt of T, electrolysis (ethyne). 32. Passed through hot Cu–tube (benzene). 33. O_3 – Zn/H_2O, Ozonolysis (U, ethanedial). 34. Anh. $AlCl_3$ – $ClCOCH_3$ (V, acetophenone), 35. Anh. $AlCl_3$, succinic anhydride (W, 4-Oxo-4-Phenylbutanoic acid). 36. D_2SO_4 (X, benzene-d). 37. $MeO_2CC \equiv CCO_2Me$, hν (Y, 1,2,-dicarbomethoxycycloocta-1,3,5,7-tetraene). 38. Me_2SO_4-dil.·NaOH (Z, anisole).

(i) EtOH, Δ; (ii) H_3PO_2; Δ; (iii) [HOH], Δ; (iv) KI; (v) Zn–conc., HCl; (vi) Cu_2X_2–HX, Δ; (vii) KCN, Cu; (viii) Na_3AsO_3, $CuSO_4$; (ix) HBF_4; (x) Cu, Δ; (xi) Δ; (xii) $NaNO_2$, Cu.

[1.9] PREPARATION OF REAGENTS

[A] Inorganic acids and ammonium hydroxide and acetic acid (of approximate strength in normality)

Sl. No.	Form. (N)	Purity (%)	Relative density	Volume + H_2O (in ml) →	Mixed and volume made upto 1L.	Will make dil. soln. of approx. Normality (N)
1.	Gl. HOAc (17)	– 99.5	– 1.05	– (290 + H_2O)	– do –	HOAc (5N)
2.	Conc. HCL (12)	– 38	– 1.19	– (420 + H_2O)	– do –	HCl (5 N)
3.	Conc. HNO_3 (16)	– 70	– 1.42	– (300 + H_2O)	– do –	HNO_3 (5 N)
4.	Conc. H_2SO_4 (36)	– 96	– 1.84	– (140 + H_2O)	– do –	H_2SO_4 (5 N)
5.	Conc. H_3PO_4 (45)	– 85	– 1.69	– (190 + H_2O)	– do –	H_3PO_4 (5 N)
6.	Liquor NH_4OH (17)	–	0.888	(340 + H_2O)	– do –	NH_4OH (5 N)
7.	SO_2 gas passed into water at room temp. to furnish sat. soln. (0.33N)					

[B] Reagents and Solutions

1. *Barfoed's Reagent (or solution)*

5g Cu $(OAc)_2$ in 100 ml of H_2O + 1 ml of gl. HOAc → Solution (Reagent).

[Used to detect glucose or similar easily oxidisable sugars. Ppt. of red Cu_2O is the positive test of such carbohydrate].

2. *Baeyer's Reagent*

Dilute soln. of $KMnO_4$ (2%), addition of dil. H_2SO_4 (few drops) is recommended. [Used to detect unsaturation or easily oxidisable functional group (s) in the organic compound. The purple colour of $KMnO_4$ is discharged. It is advisable to perform the blank test simultaneously].

3. *Benedict's Reagent (or solution)*

17.3 g $CuSO_4 \cdot 5\ H_2O$ + 100 ml of H_2O → + dil. H_2SO_4 (few drops) dissolved (soln. A).

173 g Na – Citrate + 800 ml of H_2O + 100 g anhyd. $NaCO_3$ – (soln. B)

Just before use, soln. A (1 vol.) + soln. B (1 vol) → Benedict's soln.

[Used to detect glucose or reducing sugars through formation of red Cu_2O].

4. *Brady's Reagent*

2 g of 2,4-Dinitrophenylhydrazine (2,4-DNPh) in 100 ml of pure MeOH + 4 ml of conc. H_2SO_4, stirred will, warmed, then cooled, and filtered. The clear filtrate (pale yellow soln.) is the reagent. Also 2,4-DNPh (0.25 g) in HCl (42 ml), and water (50 ml) warmed and filtered and diluted to 250 ml with H_2O.

[Used to detect ketones and aldehydes through formation of yellow, orange or red ppt. of 2,4-dinitrophenylhydrazone (2,4-DNP) and to prepare 2,4-DNP derivatives.]

5. *Bromine Water*

11 ml of Br_2 + 100 ml of H_2O stored in glass stoppered bottle.

[Used to detect unsaturation, a few aromatic amines and phenols. Useful for preparation of bromo-derivatives].

6. *Caro's Acid*

This is H_2SO_5 (monopersulphuric acid).

[Used to oxidise some special functional groups, e.g., $C_6H_4(NO_2)^4(NH_2)^1$ $\xrightarrow{H_2SO_5}$ $C_6H_4\ (NO_2)^4\ (NO)^1$ (nitroso compound)].

7. *Ceric Ammonium Nitrate (CAN)* $(NH_4)_2[Ce\ (NO_3)_6]$.

Dilute aqueous soln. is made (light yellow) just before use (very little CAN will be needed).

[Used to identify alcoholic OH gr. which turns the solution light brown. Some other groups may interfere. Matching is recommended].

8. *Chlorine Water*

Cl_2 generated from $KMnO_4$ + Conc. HCl (*caution*) is passed into H_2O until a saturated soln. is obtained. Freshly prepared solution is used.

[Used for identification of $I^{\ominus}$ and $Br^{\ominus}$ during element detection test (Lassaigne's Na Fusion test) and sometimes for oxidation purpose].

9. *Dam's Solution*

Mixture of glacial HOAc + Pyridine + Br_2 + conc. H_2SO_4.

[Used to determine the degree of unsaturation in oils and fats].

10. *Denige's Reagent*

5 g of HgO+ 20 ml of conc. $H_2SO_4 \xrightarrow[\text{to make}]{+H_2O}$ 100 ml → cooled and filtered clear soln. (Denige's Reagent).

[Used to indentify citric acid which gives a white ppt. (I)].

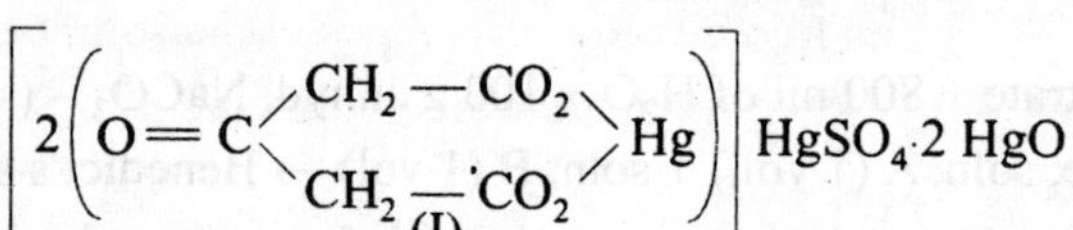

(I)

11. *Fehling's Solution*

(a) 34.6 g of $CuSO_4 \cdot 5H_2O$ in 500 ml of H_2O + few drops of H_2SO_4 – (soln. A or I).

(b) 175 g of Sodium Potassium Tartrate (Rochelle salt) + 70 g of NaOH in 500 ml of H_2O → (Soln. B or II).

1 vol. of A (i.e. I) and 1 vol. of B (i.e. II) are mixed just before use.

[Used to estimate (and identify) glucose and similar easily oxidisable organic compounds. A red Cu_2O is precipitated when the reaction is positive].

12. *Fenton's Reagent*

A mixture of H_2O_2 and $FeSO_4$.

[Used to oxidise some special alcohols to aldehydes, e.g. oxidant in carbohydrate chemistry and $HOCH_2 - CH_2 - OH \xrightarrow[\text{Reagent}]{\text{Fenton's}} HOCH_2 - CHO$].

13. *Girard's Reagent*

(a) Trimethylaminoacetohydrazide chloride = Girard T.

(b) Pyridiniumacetohydrazide chloride = Girard P.

[Used to separate oxo compounds through water soluble derivatives by filtration from water insoluble impurities. The terminal NH_2 group reacts with >C=O group to yield >C = N – derivatives],

$$Me_3N + Cl - CH_2 - CO_2C_2H_5 + N_2H_4 \rightarrow Me_3\overset{+}{N} - CH_2 - CO - NHNH_2]\overset{\ominus}{Cl} = \text{Girard T.}$$

$$C_5H_5N + ClCH_2 - CO_2C_2H_5 + N_2H_4 \rightarrow C_6H_5\ N - CH_2\ \ CONHNH_2]\overset{\ominus}{Cl} = \text{Girard P.}$$

(Pyridine).

14. *Lemieux Reagent*

Aqueous soln. of sodium periodate, $NaIO_4$, mixed with small amount of dil. $KMnO_4$ soln.

[Used to break >C = C< into two oxo compounds thus locating the position of the alkene bond. Terminal CH_2 group is converted into methanal]. Thus,

$$\begin{matrix} R_1 \\ R_2 \end{matrix}\!\!>C=C<\!\!\begin{matrix} R_3 \\ R_4 \end{matrix} \xrightarrow{\text{L.R.}} \begin{matrix} R_1 \\ R_2 \end{matrix}\!\!>C=O + \begin{matrix} R_3 \\ R_4 \end{matrix}\!\!>C=O$$

15. *Lucas Reagent*

15 g of anhydrous $ZnCl_2$ + 15 ml of conc. HCl → Lucas Reagent.

[Used to differentiate the 1°, 2° and 3° alcohols. Thus,

(a) 1° alcohols (C_n – n less than 6)—remains clear with Lucas Reagent.

(b) 2° alcohols—gives fine oily ppt. very slowly (the soln. becomes cloudy).

(c) 3° alcohols—gives two separate layers more quickly].

16. *Mayer's Reagent*

13.5 g of $HgCl_2$ (poison) in 940 ml of H_2O + 50 g of KI → Mayer's Reagent. - [Used to detect alkaloids is solutions].

17. *Molisch's Reagent*

10% alcoholic solution of α-naphthol.

[Used to detect several carbohydrates].

18. *Nessler's Reagent*

50 g of KI in 35 ml of pure H_2O + sat. $HgCl_2$ soln. ⟶ light ppt.

↓ 400 ml KOH or NaOH (9N)

Volume made 1L.
(Nessler's Reagent)

[Used to detect traces of ammonia (brown colouration or ppt. depending on the quantity of ammonia)].

19. *Pavy's solution*

8.361 g $CuSO_4 \cdot 5H_2O$ dissolved in H_2O $\xrightarrow{\text{made}}$ 1L. (soln. I).

40.8 g Rochelle salt + 40.8 g KOH + 600 ml NH_4OH (sp. gr. 0.880) → 1 L (soln. II).

Soln. I and Soln. II are mixed (1:1, v/v) just before use.

[Used as a substitute of Fehling's solution in sugar estimation].

20. *Schiff's Reagent*

Fuchsine (i.e. rosaniline 0.5 g) hydrochloride in 250 ml, of H_2O was taken then SO_2 is passed through this cold soln. until the red colour of the dye becomes colourless. The fresh soln. is used.

[Used to detect aldehydes which restore the red colour of the reagent. Ketones cannot restore the red colour thus useful to differentiate ketones from aldehydes].

21. *Schweitzer's Reagent*

$CuSO_4 \cdot 5H_2O$ dissolved in NH_4OH (dil. soln.).

[Used to dissolve cellulose].

22. *Styphnic Acid*

(2,4,6-trinitro-3-hydroxyphenol).

[Used to prepare styphnates of aromatic hydrocarbons (like picrates)].

23. *Tollen's Reagent*

A mixture of 0.1 (N) $AgNO_3$ aqueous soln. (5 ml) and 1 drop of dil. NaOH soln. is made. Very faint black ppt. is formed which is dissolved by dropwise addition of dil. NH_4OH avoiding excess. This is the reagent.

[Used to oxidise readily oxidisable organic compounds, specially glucose and similar compounds holding aldehyde group. The reagent $[Ag(NH_3)_2]^{\oplus}$ is reduced to metallic silver which is either precipitated as black powder (if the flask is shaken) or as metallic silver mirror (when the flask is kept still). The reagent is explosive when dry, so it should be decomposed (by dil. HNO_3) as soon as the experiment is over].

[C] Catalysts which are Commonly used

1. *Adams catalyst* Platinum oxide is reduced with H_2 to a mixture of Pt–PtO which is used as catalyst to reduce at room temperature, the $C = C$ and $C \equiv C$ into saturated C–C bond. The catalyst can be made with charcoal (as Pt/C).
2. Palladium (finely powdered) can be used as catalyst to reduce at room temp. the unsaturated compounds into saturated compounds in H_2 atmosphere.
3. Finely divided nickel (Sabatier–Senderens catalyst) reduces the alkene derivatives into alkane derivative in presence of H_2 at a temp. 200–300°C.
4. *Raney nickel* Ni–Al alloy is digested with NaOH resulting in finely divided nickel which is washed and stored under H_2O or EtOH. It can reduce unsaturated compounds (sometimes also at room temp.) into saturated compounds in presence of H_2. It is considered better than nickel as catalyst.
5. *Lindlar catalyst* Mixture of Pd (finely powdered), powdered $CaCO_3$ and Pb $(OAc)_2$ is the catalyst which is used to reduce alkynes into alkenes. Alkenes are not reduced to alkanes. Another variety of catalyst (Cram's catalyst) is derived from finely divided Pd on $BaSO_4$ "poisoned" (i.e. power of reduction of Pd is lowered) with quinoline. This is equally suitable like Lindlar catalyst where the lead acetate acts as a poison to reduce the power of Pd.

Thus:

6. *Ziegler–Natta catalyst* The complex formed of triethyl aluminium (Et_3Al) and titanium chloride ($TiCl_4$) is the catalyst used to prepare specific type of polymers.

1.10 A FEW LESS COMMON REAGENTS

1. *Bang's Reagent:* In H_2O (700 ml) added K_2CO_3 (100 g), then KCl (66 g), finally $KHCO_3$ (160 g), $CuSO_4$ (4.4 g) added, CO_2 evolution was complete, then diluted to 1 L. Left for 24 hrs. (air excluded). This soln. (300 ml) was diluted with KCl soln. (saturated) to 1 L and after 24 hrs. used to determine glucose.
2. *Esbach's Reagent:* Picric acid (10 g) in H_2O, added citric acid (20 g) and volume made 1 L [used to determine protein].
3. *Hager's Reagent:* Picric acid (1 g) in H_2O (100 ml) [used to detect alkaloids].
4. *Hanu's Reagent:* Dissolved I_2 (13.2 g) in gl. HOAc (1 L) and Br_2 added to double the halogen content [used to determine iodine content].
5. *Meyer's Reagent:* Dissolved $HgCl_2$ (1.358 g) in H_2O (60 ml) and poured onto a soln. of KI (5 g) in H_2O (10 ml), then diluted to 100 ml [detection of alkaloids].
6. *Nylander's Solution:* Dissolved bismuth subnitrate (20 g) in 8% NaOH solution (1 L) and Rochelle salt (40 g) added, cooled and filtered [used to detect carbohydrates].
7. *Potassium pyrogallate:* Dissolved KOH (50 g) in H_2O (100 ml) and pyrogallol (5 g) added to it. [The soln. is used to absorb O_2 from gas mixture containg less than 28% of oxygen].
8. *Wagner's Reagent:* Dissolved I_2 (2 g) and KI (6 g) in H_2O (100 ml) [used to detect alkaloids].
9. *Wij's iodine monochloride solution:* Dissolved I_2 (13 g) in gl. HOAc (1 L) and dry Cl_2 gas is passed through 975 ml of this soln. (25 ml is set aside) until the solution became colourless. To this the remaining 25 ml of the iodine in HOAc added and stored. [The solution is used to determine iodine number within one month of its preparation.]

1.11 SOME IMPORTANT TERMS USED IN PREPARATIVE ORGANIC CHEMISTRY

It is necessary for the beginners to be well-acquainted with the following terms.

(i) *Substrate (or starting material):* This is the organic molecule on which the reagent acts. Reagents are generally inexpensive and readily available in the market and mostly inorganic in nature. Generally, the substrate is made in the laboratory, more expensive, time consuming to prepare and never taken in excess compared with the reagent. The yield of the product is calculated on the basis of the substrate since its amount is generally lower than that of the reagent(s). The following examples will illustrate these points.

(a) $$\underset{\substack{\text{(Substrate)}\\ \text{(Organic)}}}{2\,C_2H_5OH} \xrightarrow[\uparrow\downarrow\ \text{(Inorganic)}]{\text{Reagent Conc. } H_2SO_4} \underset{\substack{\text{(Product)}\\ \text{(ether)}}}{C_2H_5\ O\ C_2H_5} + H_2O$$

In reaction between organic and inorganic compounds the organic compound is considered as substrate.

(b) $$\underset{\substack{\text{Inexpensive,}\\ \text{purchased}\\ \text{(used in smaller}\\ \text{quantity)}}}{R\text{–}COOH} \xrightarrow[+\ \text{Reagent conc. } H_2SO_4\ \uparrow\downarrow]{HOC_2H_5\ \text{(purchased), used in excess}} \underset{\text{Product (ester)}}{R\text{–}CO_2\ C_2H_5 + H_2O}$$

The compound in lower amount is taken as substrate, and the yield of the product is calculated on the basis of the substrate.

(c) $$\underset{\substack{\text{Readily}\\ \text{available,}\\ \text{inexpensive}\\ \text{(substrate)}}}{4\ R\text{–}CO\text{–}R'} + \underset{\substack{\text{Readily}\\ \text{available,}\\ \text{but expensive}}}{LiAlH_4} \xrightarrow[\text{room temp.}]{\text{anhydrous } Et_2O} \underset{\text{(Product)}}{4R\text{–}CH(OH)\text{–}R'}$$

The organic compound is used as substrate, and the inorganic reagent is used only in slight excess (ca 1.1 molar ratio), since this is expensive.

(ii) *Product:* The final-molecule of a chemical reaction is the product which is aimed at, often called the desired material.

(iii) *Intermediate product:* Sometimes before getting the desired product an intermediate product is isolable which can be finally converted into the desired product. This intermediate product, if unstable, will be converted into the product very fast thus escaping the isolation, but its formation can be established by various instrumental and chemical processes. Thus

$$\underset{\text{(substrate)}}{R\text{–}CO\text{–}NH_2} \xrightarrow[\text{Warm (condition)}]{Br_2 + KOH\ \text{(Reagents)}} \underset{\substack{\text{(Intermediate)}\\ \text{(Isolable)}}}{[R\text{–}CO\text{–}NHBr]} \xrightarrow[\text{Warm}]{KOH} \underset{\text{(Product)}}{R\text{–}NH_2.}$$

(iv) *Transition state:* To explain the mechanism and pathway of certain reaction it becomes necessary to assume (sometimes) the participation of an imaginary (not isolable like the intermediate product) step which helps the understanding of the reaction, particulary with respect to stereochemistry of the prodnet.

The following example is illustrative,

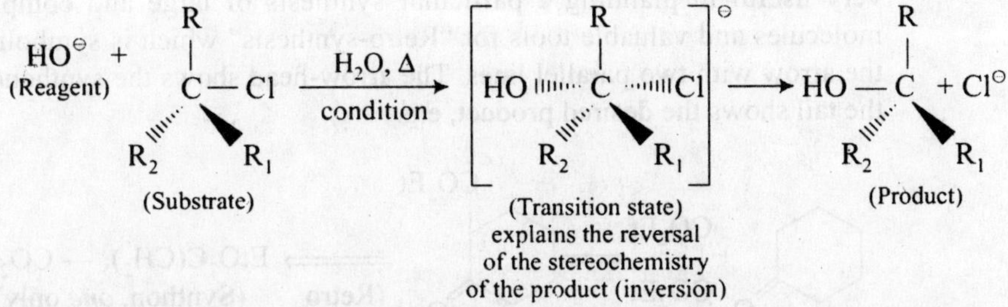

(v) *Byproduct:* During a chemical reaction in addition to the desired product other product(s) may be formed which is well defined pure material and may be converted into other useful chemicals (product). The amount of the byproduct is lower than the desired product but may pose some problems for the isolation of the desired product in pure state. Thus, during the preparation of saccharin the compound I is required as the product but II is also formed simultaneously and is termed the byproduct.

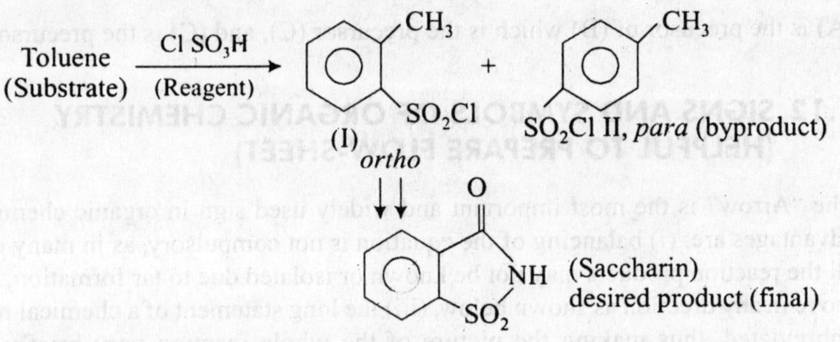

(vi) *Tarry material:* In many chemical reactions in addition to the desired product a tar-like product is also formed and the isolation of the pure desired product becomes very difficult. Sometimes, the reaction leads to only tarry product if the reaction condition is not properly maintained.

e.g. Naphthalene (substrate) $\xrightarrow[\Delta]{KBrO_3/H^{\oplus}}$ Only tarry material, no desired 1,2 (–) or 1,4-naphthaquinone can be isolated.

(vii) *Synthon:* When easily anailable, simple and small organic molecules combine together to form the larger, complex and desired molecule, then those simple molecules are called synthons. In some cases, a single compound may be considered as synthon if it is more simple structurally. These synthons are very useful in planning a particular synthesis of large and complicated molecules and valuable tools for "Retro-synthesis" which is symbolised by the arrow with two parallel lines. The arrow-head shows the synthon(s) and the tail shows the desired product, e.g.

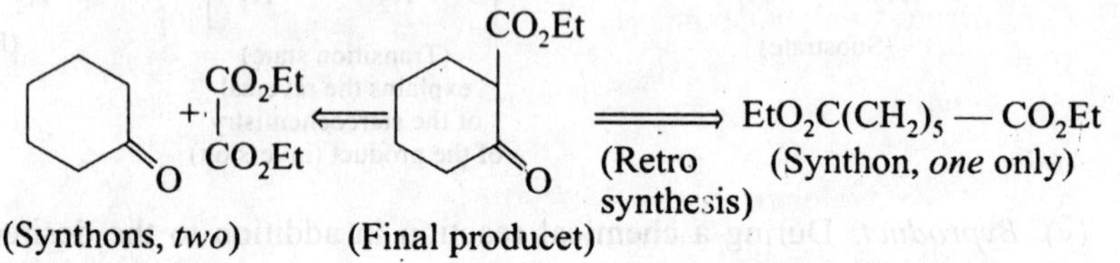

(viii) *Precursor* (just previous compound in a series of steps of a reaction).

Example

$$C_6H_5CN\ (A) \xrightarrow[\Delta]{H_3O^{\oplus}} C_6H_5CO_2H\ (B) \xrightarrow{PCl_5} C_6H_5\ COCl\ (C) \xrightarrow{NH_3} C_6H_5\text{–}CO\text{–}NH_2 (D)$$

(A) is the precusor of (B) which is the precursor (C), and (C) is the precursor of (D).

1.12 SIGNS AND SYMBOLS OF ORGANIC CHEMISTRY (HELPFUL TO PREPARE FLOW-SHEET)

The "Arrow" is the most important and widely used sign in organic chemistry. The advantages are: (i) balancing of the equation is not compulsory, as in many occasions all the reaction products may not be known or isolated due to tar formation, (ii) it can move in any drection as shown below, (iii) the long statement of a chemical reaction is abbreviated, thus making the picture of the whole reaction very briefly depicted making it more easy and quick to understand and remember, (iv) it saves time and space and (v) it helps to compare different methods of preparation of the same product more easily. The points are illustrated in the following examples.

(a) Presentation of the arrow sign:

Tail side	Body of the arrow	Head side
Substrate and other reagents. Their wts (in gm), moles, formulas and other noteworthy properties, and warning (if any).	Condition of the reaction, viz. catalysts, reagents, ⟶ concentration, time, temperature, etc.	Desired product, byproduct, formula (name), yield, wt. (gm; %), mole, noteworthy properties, colour, m.p. or b.p. odour, etc. and warning (if any)

e.g. (i)

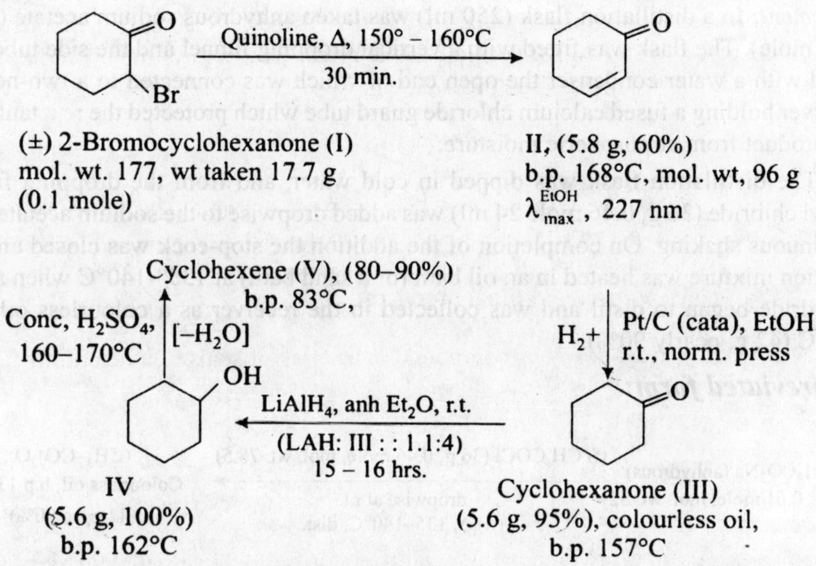

NOTES:

- **The arrow can propagate in any direction.**
- **If the full names of the compounds are given, the corresponding identifying numerals (usually Roman but sometimes Arabic) are given in parantheses just after the names.**
- **Problem solving becomes easier through the "flow-sheet" using the arrow because sometimes the text becomes complicated.**
- **But the chemical properties (like poisonous, explosives, etc.) are not mentioned, moreover the method of isolation of the products (i.e. work-up procedure) is never outlined.**
- **Deseription of the reaction steps from which above flow-sheet (i) made. Racemic 2-bromocyclohexanone (I, 17.7 g) with excess of quinoline at 150–160°C for 30 mins. furnished the desired α, β-unsaturated ketone (II), b.p. 168°C (5.8 g, 60%), EtOH λ_{max} 227 nm. (nature of the apparatus, detail reaction condition and work-up procedure, etc. are missing).**

When compound II (5.8 g) was dissolved in ethanol and hydrogenated at room temperature and atmospheric pressure with Pt/C catalyst, the product was cyclohexanone (III, 5.6 g, ca 95%) obtained as colourless oil, b.p. 157°C.

The compound III (5.6 g) was reduced with LiALH$_4$ in anhydrous diethyl ether at room temperature to cyclohexanol (IV) in almost quantitative yield, b.p. 162°C. The molar ratio of the reductant and the ketone III was 1.1:4, showing just a little excess of the reductant was used. The alcohol IV on dehydration by heating with conc. H_2SO_4 at 160-170°C furnished the desired cyclohexene as colourless oil, b.p. 83°C in 80–90% yield.

e.g. (ii) *Abbreviating a chemical reaction through arrow sign:*

Statement: In a distillation flask (250 ml) was taken anhydrous sodium acetate (50 g, 0.61 mole). The flask was fitted with a vertical dropping funnel and the side tube was fitted with a water condenser the open end of which was connected to a two-necked receiver holding a fused calcium chloride guard tube which protected the reactants and the product from atmospheric moisture.

The distillation flask was dipped in cold water, and from the dropping funnel acetyl chloride (36 g, 0.46 mole 24 ml) was added dropwise to the sodium acetate with continuous shaking. On completion of the addition the stop-cock was closed and the reaction mixture was heated in an oil bath (or a sand bath) at 135°-140°C when acetic anhydride began to distil and was collected in the receiver as a colourless oil, b.p. 138°C (42 g, nearly 90%).

***Abbreviated form*:**

$$\underset{\text{(50 g, 0.61 mole, mol. wt. 82)}}{CH_3CO_2Na \text{ (anhydrous)}} \xrightarrow[\text{(ii) } \Delta,\ 135\text{–}140°C,\ \text{dist.}]{\text{(i) } CH_3COCl \text{ (36 g, 0.46 mole, mol. wt. 78.5) dropwise at r.t.}} \underset{\substack{\text{Colourless oil; b.p 138°C} \\ \text{(42 g, ca 90\%)}}}{(CH_3\text{–}CO)_2O}$$

(iii) *Writing the flow-sheet of the problem makes it simpler. The arrowhead is directed towards the products.*

e.g. *Problem:* An organic compound (A) on treatment with metallic Na evolves H_2 and gives (B). Again, when (A) is reacted with iodine in presence of dilute NaOH at 70°C then (C) is formed which on heating with dilute NaOH yields sodium formate. In separate experiment (D) is obtained by heating (C) with powdered Ag. Now (D) reacts with dil. H_2SO_4 in presence of $Hg^{2\oplus}$ as catalyst to give (E) which on treatment with I_2 and dil. NaOH affords (C). Moreover, when (A) is oxidised with dil. $KMnO_4$ and dil. H_2SO_4 then the acid (F) is formed. The Na–salt of (F) is mixed with soda–lime and strongly heated to produce methane. Identify all the compounds marked as (A) through (F).

Solution: Attempted through the following simple flow-sheet (using arrows)

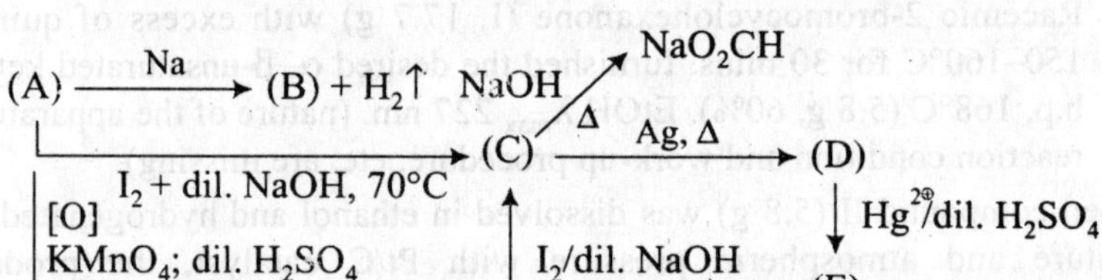

1.13 CALCULATION OF THE PERCENTAGE YIELD OF THE PRODUCT

(i) % yield of the product = $\dfrac{\text{wt (in g) of product obtained}}{\text{wt (in g) of product calculated}} \times 100$

(ii) Calculation must be done on the basis of the reactant having minimum quantity (in moles). Thus, in e.g. (ii) the yield is 90% on the basis of AcCl while it is only 68% on the basis of NaOAc.

(iii) Material balance is necessary when possible.

e.g. *% yield calculation and material balance:*

$$\underset{17\text{ g}}{C_6H_5-CO_2H} \xrightarrow[\uparrow\downarrow\ 16\text{ hrs}]{\text{EtOH (excess), conc. } H_2SO_4} \underset{14.5\text{ g}}{C_6H_5-CO_2C_2H_5} + \underset{3.2\text{ g}}{C_6H_5COOH}$$

Calculations:

(i) EtOH is in excess, so the yield is calculated on $C_6H_5CO_2H$.

(ii) The mol. wt. of $C_6H_5CO_2H$ is 122, and the mol. wt. of $C_6H_5CO_2C_2H_5$ is 150.

(iii) Recovery of $C_6H_5CO_2H$ is 3.2 g. So ester is formed from (17.0–3.2) g of $C_6H_5CO_2H$.

(iv) 122 g of $C_6H_5CO_2H$ when gives 150 g of $C_6H_5CO_2C_2H_5$, the yield is 100%.

So (17.0 – 3.2) g = 13.8 g of $C_6H_5CO_2H$ should give $\dfrac{150 \times 13.8}{122}$ g = 16.9 g of $C_6H_5CO_2C_2H_5$.

(v) 16.9 g ester is the calculated and 14.5 g is the actual yields, respectively.

(vi) Therefore, percentage yield in this reaction is $\dfrac{14.5\text{ g} \times 100}{16.9\text{ g}}$ = 85.8% based on recovery.

(vii) Material balance; Recovery of $C_6H_5CO_2H$ = 3.2 g

14.5 g of $C_6H_5CO_2C_2H_5$ contains $C_6H_5CO_2H$ = 11.8 g

Total 15.0 g

Therefore, loss of $C_6H_5CO_2H$ during work-up = (17.0 – 15.0) g = 2 g.

1.14 FEW MORE SIGNS USED IN ORGANIC CHEMISTRY (USEFUL FOR THE BEGINNERS):

(i) $\longrightarrow$ Sign for *inductive effect* (also called inductive static, Is).

e.g. A $\longrightarrow$ B

A on the tail side, less electron attracting compared with H (e.g. CH_3, $-CR_3$ etc.)

B on the arrow head side B is more electron attracting compared with H, (e.g. Cl, CN etc.)

(ii)a. —+—→ Sign for dipole moment; expressed as μ, unit D (Debye, $= 10^{-18}$ esu.).

e.g. A —+—→ B

Tail side is δ⊕, less electron attracting | Head side is δ⊖, more electron attracting

- *Electronegativity order:* F (4.0), O(3.5), Cl/N (3.0), Br (2.8), S/C/I (2.5), H (2.1) Si (1.75).
- *Identification of stereoisomers through resultant* μ (= μ_R):

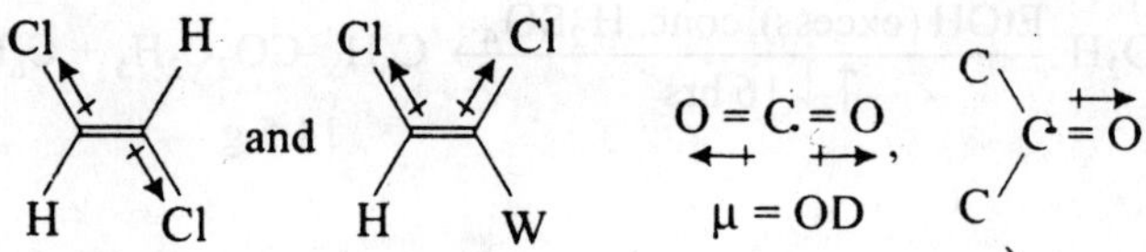

μ = OD, *trans* and μ = 2.95D *cis*; O = C = O, μ = OD; μ_R > OD.

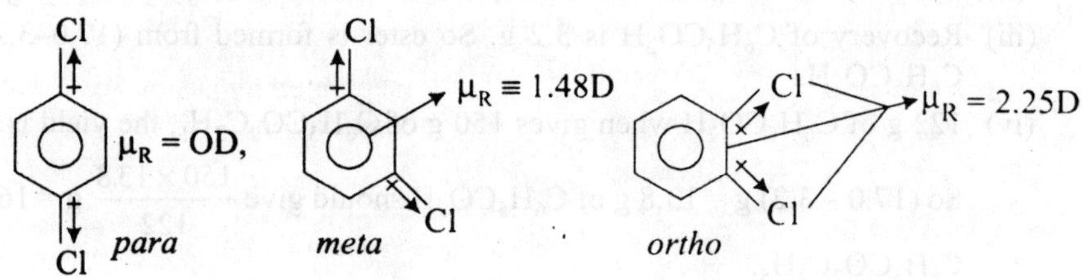

μ_R = OD, *para*; $\mu_R \equiv 1.48D$ *meta*; $\mu_R = 2.25D$ *ortho*

$\mu_R^2 = \mu_1^2 + \mu_2^2 + 2\,\mu_1\,\mu_2 \cos\theta$ (θ = angle between μ_1 and μ_2; $\cos 60° = \frac{1}{2}$, $\cos 120° = -\frac{1}{2}$, $\cos 180° = 1$)

[Identification of *o*, *m* & *p*-isomers of dichlorobenzene].

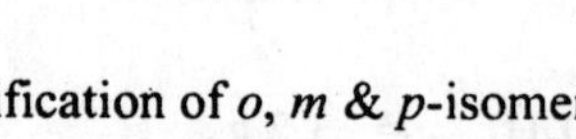

$\mu_2 = 1.5D$, $\mu_1 = 4D$. So $\mu_R = 4D - 1.5D = 2.5D$ (for *p*-isomer).

(ii)b. —×—→, No reaction e.g. $R–CH_2–OH \xleftarrow{HNO_2} R–CH_2–NH_2 \xrightarrow[\times]{HNO_2} R–CH_2–\overset{\oplus}{N}_2\,X^{\ominus}$

(iii) ⇶, Multistep reactions from substrate to the desired product.

e.g.

(Benzene) ⇶ (Salicylic acid)

(iv) H----, Hydrogen bonding (between two electronegative atoms thus making significant change in the physical properties).

e.g. Boiling points are proportional to the molecular weights. But H_2O (mol. wt. 18) b.p. 100°C while H_2S (mol. wt. 34), gas at r.t. This is due to H-bonding in water.

- C_2H_5OH (mol. wt. 46) b.p. 78°C, Me–O–Me (mol. wt. 46) b.p. –23°C, due to H-bonding in C_2H_5OH (as shown below) making the molecule as $(C_2H_5OH)_n$.

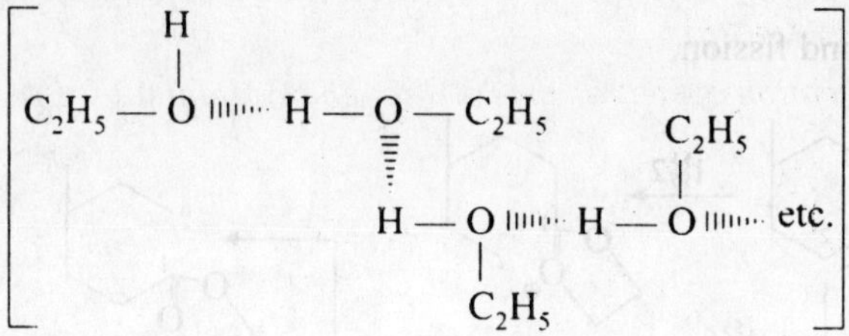

- 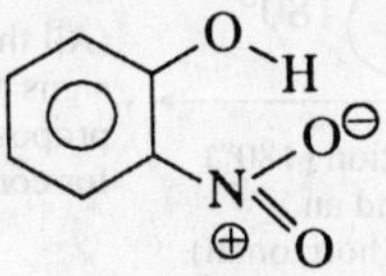 and 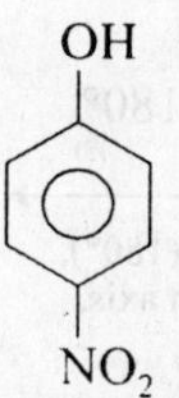 (The mixture may be separated by steam distillation)

o-Nitrophenol (steam volatile) (H-bonding) | *p*-Nitrophenol (*not* steam volatile)

(v) $\xrightarrow{h\nu}$, Photochemical reaction, h = Planck's constant = 6.625×10^{-27} erg sec and is the frequency of light.

e.g. $CO_2 + H_2O \xrightarrow{h\nu} H_2CO + O_2$; $(CH_3)_2C = O \xrightarrow{h\nu} C_2H_6 + CO$.

(vi) $\xrightarrow{\downarrow\uparrow}$, Heated under reflux.

e.g. $R{-}CN + H_2O \xrightarrow[\downarrow\uparrow,\ 20\ hrs]{conc.\ HCl\ +\ HOAc\ (glacial)} R{-}CO_2H + NH_4Cl$

(vii) $\xrightarrow{\Delta}$, Pyrolysis or thermolysis (bond breaking by heat)

e.g. $CH_3{-}COONa \xrightarrow[\Delta]{Soda-lime} CH_4 + Na_2CO_3$.

(viii) $\xrightarrow{)))}$, Sonochemical sign (reactions carried by sound energy).

e.g. $C_6H_5 \cdot CH_2{-}CO_2CH_3 \xrightarrow[Zn(BH_4)_2]{))),\ 18h} C_6H_5 - CH_2 - CO_2H$

(ix) $\xrightarrow{\circlearrowright}$, Stirring (only proposed for consideration).

e.g. $\begin{matrix} R \\ R_1 \end{matrix}\!\!>C=O \xrightarrow[\ominus]{LiALH_4,\ Ether} \begin{matrix} R \\ R_1 \end{matrix}\!\!>CH-OH$

(x) $\xrightarrow{Anode,\ -e}\cdot$ Electrochemical sign.

e.g. $CH_3-\underset{\underset{O}{\|}}{C}-\ddot{\underset{\cdot\cdot}{O}}:^{\ominus} \xrightarrow[-e]{Anode} CH_3-\underset{\underset{O}{\|}}{C}-\ddot{\underset{\cdot\cdot}{O}}\cdot \longrightarrow \dot{C}H_3 + :\ddot{O}=C=\ddot{O}:$

(xi) $\xrightarrow{x \wr\wr y}$, Bond fission,

e.g.

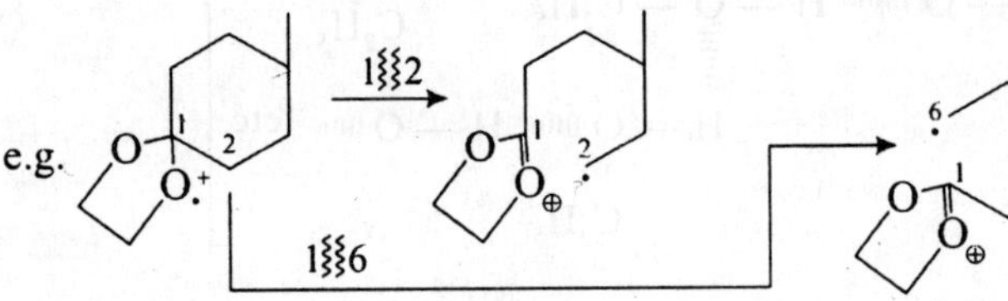

(xii) $\xrightarrow[\text{Rotation, (180°), around an axis perpendicular to the plane of the paper.}]{180°}$, $\xrightarrow[\text{Rotation (180°), around an axis, (vertical).}]{180°}$, $\xrightarrow[\text{Rotation (180°) around an axis (horizontal).}]{180°}$, (All these signs are proposed only for consideration)

(xiii) $\xrightarrow{\curvearrowright}$, Biochemical Reaction

Substrate $\xrightarrow{\text{Reagent} \curvearrowright \text{Spent reagent}}$ Product,

e.g.

$$\begin{matrix} CHO \\ | \\ (CHOH)_4 \\ | \\ CH_2OH \end{matrix} \xrightarrow{ATP \curvearrowright ADP} \begin{matrix} CHO \\ | \\ (CHOH)_4 \\ | \\ CH_2-O-PO_3^{2\ominus} \end{matrix}$$

(xiv) $\overset{\curvearrowright}{\ddot{A}-B}$, Bent arrow, known as electromeric or inductomcric abbreviated as Id (induction dynamic) effect.

This sign indicates that A is the "migration origin" from where a pair of electrons migrates to B, whereby A becomes $A^{\oplus}$ (temporarily, just before the chemical reaction) and B becomes $B^{\ominus}$. This is considered as temporary polarisation. Thus $\rangle C::C\langle \;\longleftrightarrow\; \rangle \overset{\oplus}{C}:\overset{\ominus}{C}\langle$ (not $\rangle \overset{\ominus}{C}=\overset{\oplus}{C}\langle$). Since this bent arrow involves migration of electrons (not atomic species), so it is also used to represent resonance.

e.g.

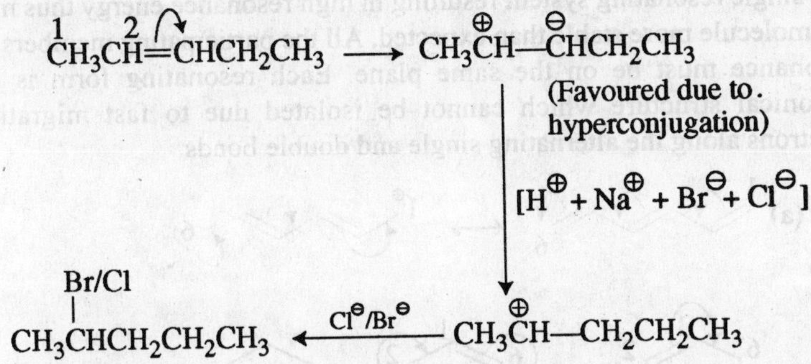

(xv) ⌒, Heavy headed arrow, (Shanon introduced, 1963, abandoned later) was used to indicate migration of *two* electrons resulting in charged centre formation.

(xvi) ⌒, Light headed arrow (Shanon introduced to indicate migration of *one*-electron only, resulting in bond cleavage and formation of free radicals). Thus,

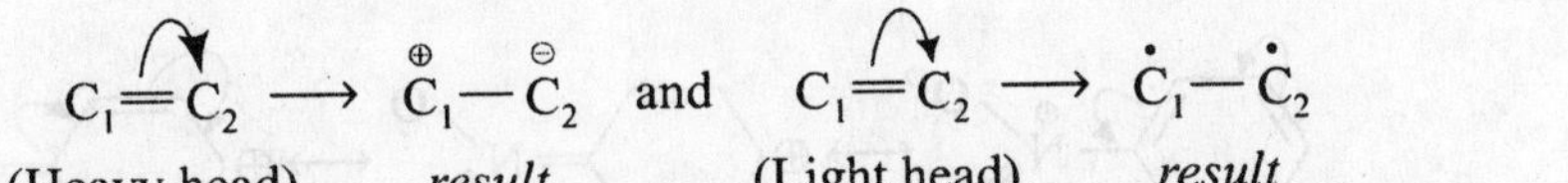

Due to obvious reason these two signs were abandoned in a short time.

(xvii) ⌒, Fish-hook arrow (Djerassi, 1964).
Used to denote migration of *one* electron resulting is generation of two free radical centres by the cleavage of one bond only. Thus,

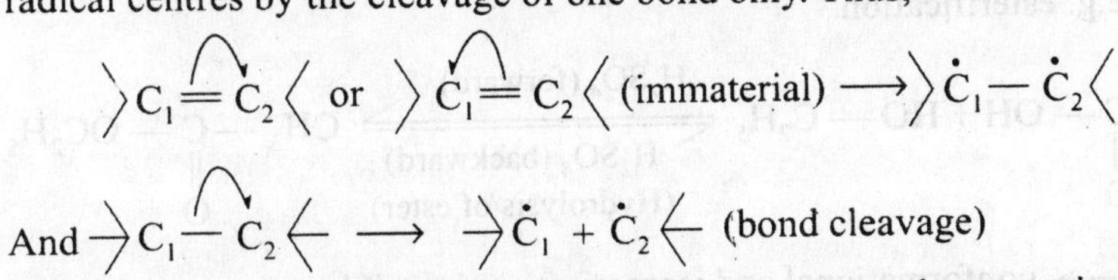

Djerassi used the original bent arrow to represent *two* electron migration.

(xiii) ——O——► or ——Θ——►, Inversion sign, used to denote the inversion of configuration of the substrate during substitution. Thus,

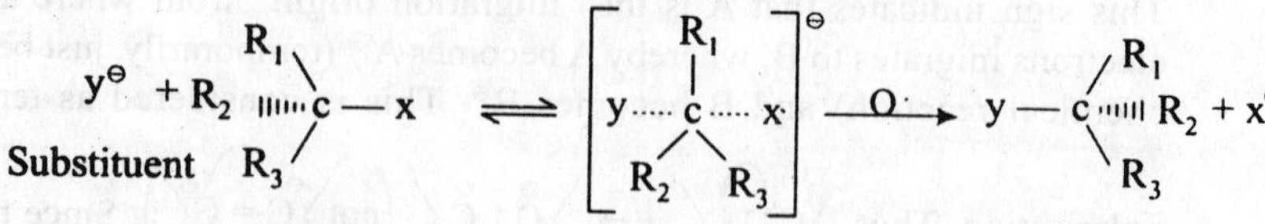

Substituent Substrate Transition state Product, (inversion)

(xix) $\longleftrightarrow$ or $\longleftrightarrow$, Resonance sign.

Very fast migration of electrons in the molecule where all the atoms take part in a single resonating system resulting in high resonance energy thus making the molecule more stable than expected. All the participating members of the resonance must be on the same plane. Each resonating form is called canonical structure which cannot be isolated due to fast migration of electrons along the alternating single and double bonds.

e.g. (a)

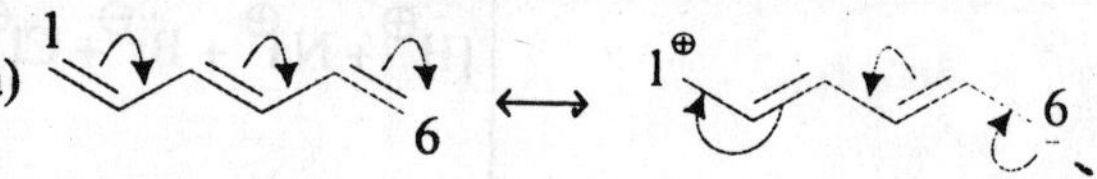

(b)

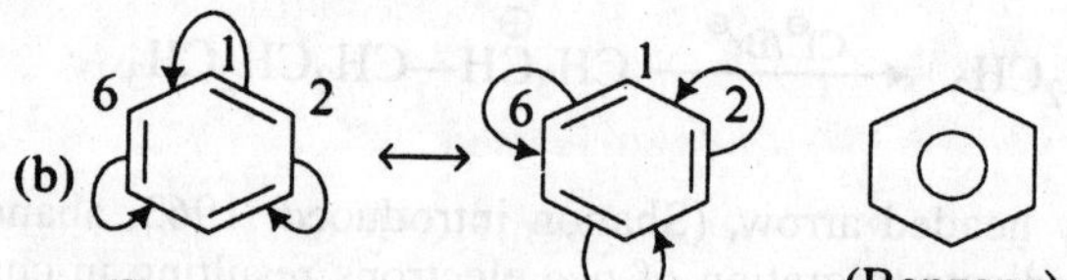

(Benzene) (Benzene) (Benzene) Robinson represented (1925)

(c)

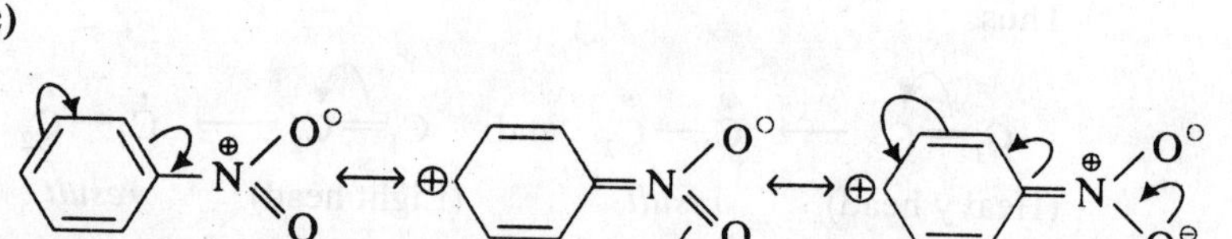

(Nitrobenzene)

(xx) $\rightleftharpoons$ or $\rightleftharpoons$ Reversible sign. There is an equilibrium between substrate and the product.

e.g. esterification

$$CH_3-\underset{\underset{O}{\|}}{C}-OH + HO-C_2H_5 \underset{\substack{H_2SO_4\,(backward)\\(Hydrolysis\ of\ ester)}}{\overset{H_2SO_4\,(forward)}{\rightleftharpoons}} CH_3-\underset{\underset{O}{\|}}{C}-OC_2H_5 + HOH$$

(xxi) $\rightleftharpoons$, conformational and isomerisational equilibrium.

e.g. 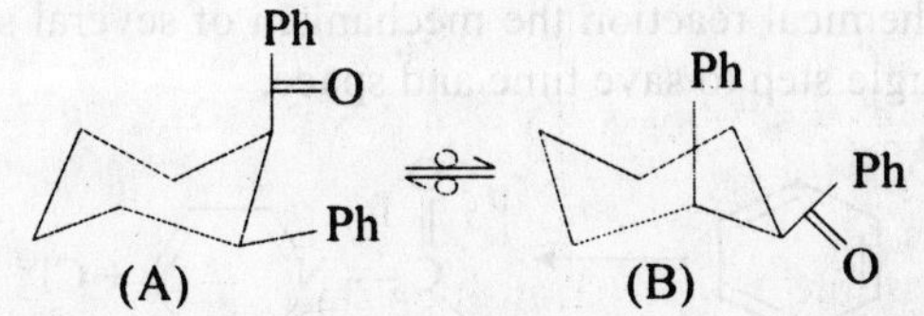

(A) (B)

(xxii) ⟹, sign of "Retro-Synthesis". The substrate lies on the arrowhead side and the desired product is on the tail side. This sign is very useful in developing a synthetic route by the synthetic organic chemists.

e.g.

(a) 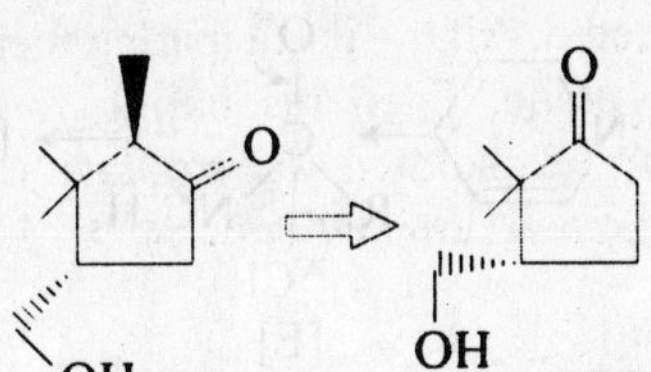

[A] [B]

(b) 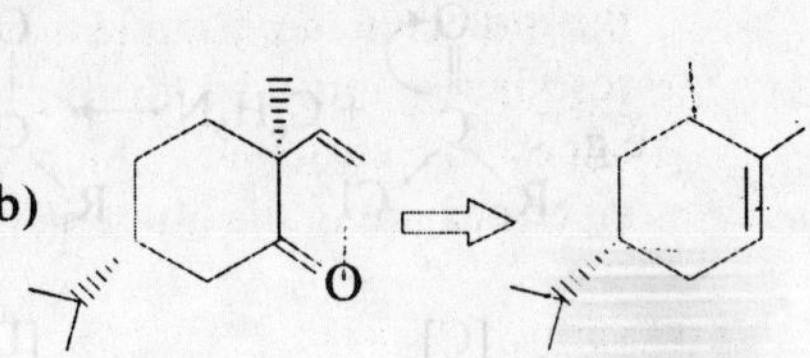

[C] [D]

(xxiii) In a reference book (W. Theilheimer) some helpful signs have been used (xxiii through xxvi).

For *addition reaction* (⇩), this sign is used.

e.g. $R{-}\underset{\underset{O}{\|}}{C}{-}H + 2H \rightarrow R{-}CH_2{-}OH$ [sign HC ⇩ OC]. This means that a CH bond is developed in place of CO bond.

(xxiv) *Rearrangement reaction* (↻) This sign is used.

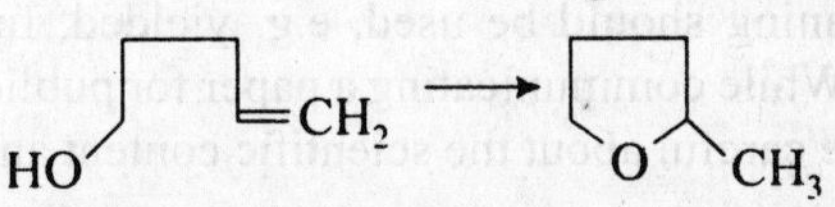

[Sign OC_{CC}↻]. This means that a CC bond is broken and a CO bond is formed.

(xxv) *Exchange reaction* (↑↓) In the following example, this sign indicates formation of a CN bond with a simultaneous loss of OC bond. (↑↓ also for reflux).

e.g. $R-C(=O)-OR + H_2N-NH_2 \rightarrow R-C(=O)-NH{\cdot}NH_2 \rightarrow$ [NC↑↓OC]

(xxvi) *Elimination reaction* [⇧] In the following example, this sign is used to indicate the formation of a CC bond and the loss of C-Hal bond.

e.g. $(Me)(H_3C)C(Br)-CH(OR)_2 \longrightarrow H_2C=C(CH_3)-CH(OR)_2$ [CC ⇧ Hal]

(xxvii) Sometimes in a chemical reaction the mechanism of several steps has been condensed in a single step to save time and space,

e.g. 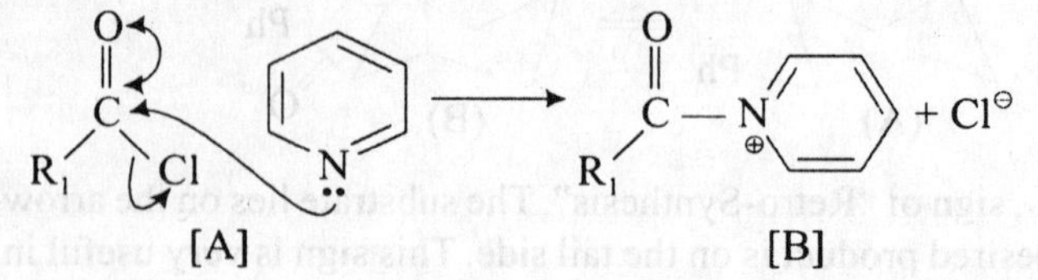

This sign indicates the result of the steps [C], [D] and [E] to form [B].

e.g. 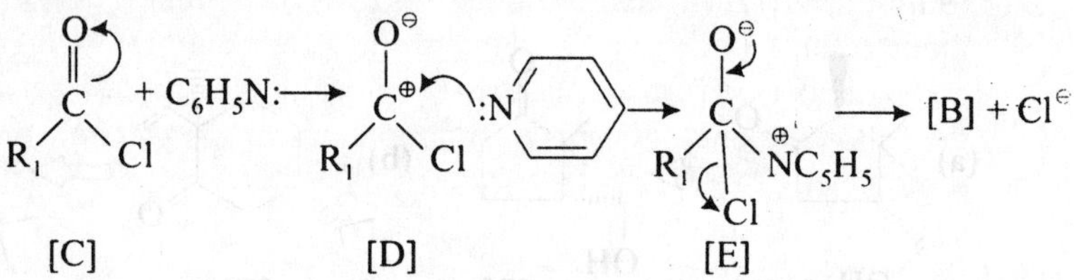

(xxviii) =, This sign means "Equals to" often called "equation sign". This is the well known sign from the very beginning of chemistry study. One thing should be clear that this should not be confused with the double bond sign of organic compound (e.g. $CH_2 = CH_2$ is not CH_2 is equal to CH_2).

1.15 PRESENTATION OF THE INVESTIGATION

(a) While repo: :ing an experiment the third person singular number and past participle tense should be used (vide 1.6). To avoid monotony of expression different words of same meaning should be used, e.g. yielded, furnished, formed, gave, produced, etc. While communicating a paper for publication in a journal the worker should be careful about the scientific content and proper presentation of the paper.

(b) The knowledge of the *distinction* between the following pairs of words should improve the communicative skill of the beginners. (Allusion, illusion); (Ascent, assent); (Apposite, opposite); (Alter, altar); [Advice (n); advise (v)]; (Affect, effect); (Council, counsel); (Complement, compliment); [Dependant (n), dependent (adj)]; (Decease, disease); (Deference, difference); (Elicit, illicit); (Eminent, imminent); (Eligible, Illegible); (Fair, fare); (farther, further); (Forward, foreword); (Equatorial, horizontal); (Licence, license); (Latter, later, letter); (Leave, live); (Loose, lose); (Nuclear, nucleus); (Plane, plain); (Patrol, petrol); (Past, paste); (President, precedent); (Principal, principle); (Peace, piece); (Pour, poor, pore); (Practice, practise); (Sight, site, cite); (Stationary, stationery); (Theoretical, theatrical); (Weak, week, wick); (Taste, test); (Wander, wonder); (Whole, hole); (Whether, weather).

(e) The knowledge of the *abbreviations* given below will improve the proper and quick understanding of the scientific literature by the beginners.

(i) A, Absolute temperature (273 K at 0°C), (ii) Å, Angstrom (10^{-8} cm), (iii) abs, absolute (100%), (iv) *ab initio* (from the beginning), (v) Ac, acetyl group, (CH_3–CO–), (vi) AcO, acetoxy (CH_3–CO–O), (vii) Alk, (alkaline), (viii) An (acetone); (ix) approx. (approximately), (x) atm (atmosphere), (xi) aq. (aqueous), (xii) b.p. (boiling point), (xiii) *bis* (two complex entity), (xiv) Bu^t (*tert.* butyl), (xv) Bn (benzyl, C_6H_5–CH_2–), (xvi) Bz (benzoyl, C_6H_5–CO–), (xvii) $\bar{c}$ (with), (xviii) *ca* (circa, approx. about, nearly), (xix) Calcd (calculated), (xx) *cf.* (*confer* = compared), (xxi) cpd (compound), (xxii) D (derivative), (xxiii) (d) = decomposition, (xxiv) *d* (Dextrorotatory), (xxv) do (ditto), (xxvi) Δ (heated or boiled), (xxvii) dil. (dilute), (xxviii) ↑↓ (heated under reflux), (xxix) e.g. (*exampli gratia* = for example), (xxx) *et al.* (*et alii* = and others), (xxxi) *etc.* (*et cetera* = so forth), (xxxii) *et seq.* (*et sequens* = and the following one), (xxxiii) *fn* (footnote), (xxxiv) *ib* (*ibid* = *ibidem* = in the same place), (xxxv) *i.e.* (*id est* = that is), (xxxvi) *in situ* (in the same site), (xxxvii) *id* (*idem* the same), (xxxviii) *iq* (*idem quod* = the same place, (xxxix) m.p. (melting point), (xL) N.B (*nota bene* = note well), (xLi) *op. cit.* (*opere citato* = in the work cited), (xLii) P.S (postscript). (xLiii) ppt. (precipitate), (xLiv) P.T.O. (Please turn over), (xLv) Ph (ϕ = C_6H_5 = Phenyl), (xLvi) *q.v.* (*quod vide* = see which,) (xLvii) t.t. (test tube), (xLviii) *u.s.* (*ut supra* = as above), (xLix) *v.i.* (*vide infra* = see below), (L) *v.s.* (*vide supra* = as above), (Li) *viz.* (videlicet = namely), (Lii), *Vs* (versus), (Liii) *v/v* (vice versa = on both ways), (Liv) v/v (volume by volume, ratio), (Lv) wt. (weight), (Lvi) X'-tract (extract), (Lvii) X's (excess); (Lviii) X'tals (crystals), (Lix) xy = xylene, (Lx) yld (yield).

(d) The beginners should carefully note the following *spelling*.

1. Abscissa, 2. access (accessory, accessible), 3. accommodation, 4. accompany, 5. accurate, 6. acknowledgment (acknowledgement), 7. across, 8. addendum (singular), addenda (plural), 9. adduct, 10. admissible, 11. afford, 12. agreeable, 13. alum, 14. amalgam, 15. ammonia, 16. apposite, 17. appropriate, 18. approximately, 19. aqueous, 20. asbestos, 21. asymmetric, symmetric, dissymmetric, 22. assure, 23. attested, 24. authentic, 25. author, 26. awkward, 27. beginning, 28. believe, 29. between, 30. burette, 31. Cal*cium*, magne*sium*, potas*sium*, stron*tium*, 32. character, 33. chlorine, cromium, chromatography, 34. cite (cf. site, sight), 35. colleague, 36. committee, (commission), 37. confirmation (cf. conformation), 38. convenience, 39. copper, 40. courtesy, 41. crystal (cf. crystallised), 42. curriculum, 43. danger, 44. decision, 45. delete, 46. descend, 47. diagnosis, 48. diamond, 49. diminish, 50. discernible, 51. discus, (cf. discussion), 52. dissent (dissension, cf. decent). 53. dissolve, 54. divide, 55. due (cf. duly), 56. dye (cf die), dyeing, 57. effervescence, 58. embarrassment, 59. endeavour, 60. equatorial, 61. exaggerate, 62. exceed, 63. exhibition, 64. false, 65. forward (cf. foreword), 66. fulfil (cf. fulfilled, fulfilment), 67 grease, 68. guarantee, 69. homogeneous, 70. hygiene,

71. eligible , 72. immense, 73. immiscible, 74. independent, 75. interference, 76. illegible, 77. itinerary, 78. lessen (cf. lesson), 79. letter (cf. later, latter), 80. liquor, 81. mercury, 82. negative, 83. noticeable, 84. nuclear (cf. nucleus). 85. occasion, 86. occur (cf. occurred). 87. palladium, 88. parallel, 89. piece, 90. phosphorus (noun) (cf. phosphorous, adj). 91. precipitate, 92. preliminary, 93. principal (cf. principle). 94. probable (cf. possible, plausible). 95. programme, 96. proceed (cf. procedure). 97. pure (cf. purity). 98. pursuit (cf. pursue), 99. pyrrole (cf. pyridine), 100. received, 101. recommend, 102. reference, 103. repeat (cf. repeatedly, repetition), 104. scissors, 105. separate, 106. serendipity, 107. sincerely, 108. soot (cf. shoot), 109. steel (cf. steal, still), 110. staggered, 111. succeed, 112. summary (cf. summer), 113. sulphur, 114. trespass, 115. trigger, 116. transferred, 117. true (cf. truly), 118. vacuum, 119. wet (cf. wait), 120. weight.

(e) The beginners should write the following *as single word*.

Almost, already, anything, cannot, everyone, into, inside, moreover, meanwhile, welcome, welfare, welkin, none, sometimes, today, tomorrow, and percentage.

(f) The following are *two words*.

All right, at least, at once, some time, well done, no one, and per cent.

Identification of Pure Organic Compounds and Other Information

2

2.1 INTRODUCTION

The study of chemistry involves theoretical and practical sections which are equally important. Practical section needs experiment, observation and inference (i.e. conclusion). It is needless to state that practical section is actually search for the truth and the motto of the worker should be "truth, only truth and nothing but truth", so the experiments should be recorded very carefully. The conclusion drawn may be wrong but the experiments and observations must be flawless.

Research workers should use better quality of notebook compared with the graduate students, each page should be numbered. The first experiment should be recorded in detail, with the flow sheet of substrate, reagents, conditions of experiment, and product(s). The description of the experiment should include (in detail).

(i) The temperature (hot, cold, reflux, etc.), solvent (density, quantity, etc.) ratio (w/w or v/v), amount of the substrate and other reactants (gm, mole, or volume and density), time required to achieve the maximum yield of the product, the shape and size of the reaction flask, etc.

(ii) Any physical change (like colour, odour, solid or liquid precipitate formation, evolution or absorption of heat, etc.), and

(iii) The collection part includes the isolation and purification of the desired product through the termination of the reaction, solvent extraction (nature of the solvent, volume and number of each extraction and in case of mixed solvents, the ratio of the components in volume).

(iv) The physical properties of the product (b.p., m.p., colour, odour, spectral data, weight in g, % yield, poisonous, etc.) and finally.

(v) Reporting the byproduct, if any, and characterisation, if possible, and precautionary measure, if needed, to avoid environment pollution, explosion, catching fire and similar hazards. The repeated experiment should refer to the number of notebook with the corresponding page mark and all experiments should be dated properly but no need of detailed description when repeated afterwards.

It is very desirable that the experiment and observation should be recorded immediately after the completion of the experiment or yet better during the progress of the reaction. Good memory of the researchers is very desirable but they should not depend on it while performing the research work. Since English is widely accepted language by the scientific community, a good knowledge of English will be very useful qualification and added advantage for the researchers. The scientific communication needs good material contents and nice presentation for its publication in reputed journals. It has been mentioned before that the experiments should be recorded in 3rd person, singular number and past participle tense.

For the graduate students the recording of the experiments and observations is much easier as they perform the routine experiments. They should record at the beginning the Roll No., Sample No., and Date of the experiments.

The identification of organic samples involves mainly the search for the functional groups in a systematic way. The philosophy of investigation in organic chemistry is to find out which functional groups are absent, and when most of the possibilities have been eliminated, the remaining one will be truth. This should be guessed and confirmed through derivative preparations.

In the following, an example of performing the experiments to identify the sample has been outlined which the graduate students may follow.

2.2 SUGGESTED EXAMPLE OF STEPWISE INVESTIGATION OF QUALITATIVE ANALYSIS

Roll No:

Sample No:

Date:

Title of the Experiment

[A] Physical characteristics:
- (a) State (solid/liquid),
- (b) m.p./b.p.,
- (c) colour,
- (d) odour,

[B] Element Detection:
- (a) Lassaigne's Sodium Fusion Test,
- (b) Beilstein Test for Halogens.

[C] Ignition Test. (How the sample burns?)

[D] Solubility Test and Classification of the Compound.

[E] Functional Group Test.

[F] Suggested Nature of the Compound from these observations.

[G] Confirmation of the m.p./b.p. through purified sample.

[H] Literature Survey and Search for Derivatives.

[I] Preparations of Apposite Derivatives.

[J] Preliminary Conclusion, Nature and Name of the Sample.

[K] Special Test of the Compound (if present).

[L] Final Conclusion:
- (a) Name of the compound
- (b) Structural formula of the compound.

(c) Any comment on the sample.

[M] Preservation of the residual sample (if pure) with proper identification mark (saving the sample and to settle the controversy, if arises, due to wrong posting).

2.3 EXAMPLE OF RECORDING IN LABORATORY NOTEBOOK (LNB)

Roll No:

Sample No:

Date:

Name of the sample analysed (Final conclusion)

[A] *Physical properties*

(a) Solid (crystalline/amorphous),

(b) liquid (colour/odour)

(c) m.p. for solid/b.p. for liquid.

(d) light/heavy.

[B] *Element Detection*

(a) Lassaigne's Sodium Fusion Test: (In this order)

S	N	Cl	Br	I	+ if present
+/–	+/–	+/–	+/–	+/–	– if absent

(b) Beilstein Test for halogens: Green flame (+), not green flame (–); when (+) halogen may be present (not sure) but when (–) halogens must be absent (sure).

[C] *Ignition Test*

(a) The sample burns with sooty flame meaning lots of carbon particles are set free signifying greater percentage of carbon and higher degree of unsaturation in the compound compared with the corresponding saturated compounds.

(b) After complete combustion if no residue is left then the sample in not a metallic salt.

[D] *Solubility Test:*

(a) In a clean t.t. small amount of the sample is taken and tested for.

Solvent	H_2O	Et_2O	5% aq. NaOH	5% aq. $NaHCO_3$	5% aq. HCl	Conc. H_2SO_4	85% H_3PO_4	Class	other org. solvents
Solubility	±	±	±	±	±	±	±	below (b)	±

(+) means sample is soluble is the solvent above in the box.
(–) means sample is insoluble.

(b) *Classification*: Each time fresh sample is taken and the letter just below the "soluble" is selected and given under "class" in the box of [D(a)]. The compounds belonging to this letter should be mentioned after the classification test. "Soluble" path should be followed.

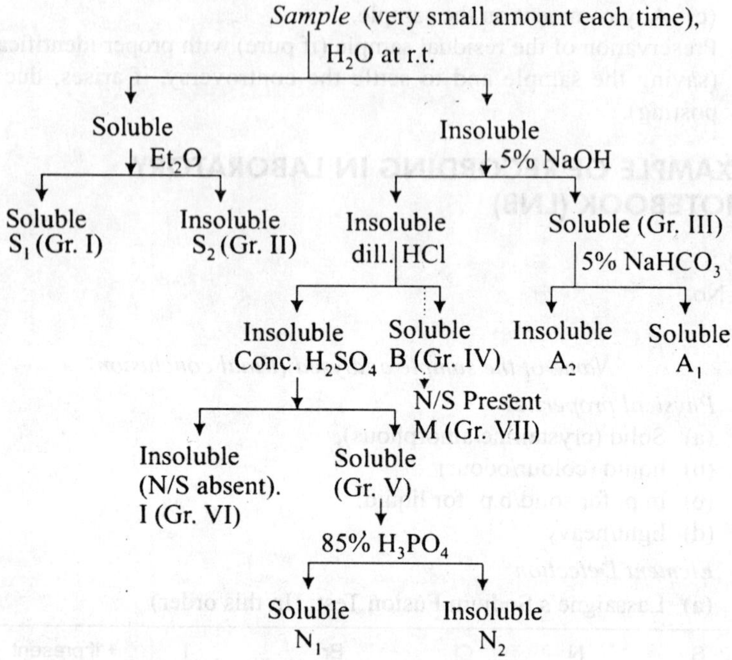

(c) *Compounds belonging to:* (List should go into the LNB).

S_1 = Low molecular weight alcohols, ketones, aldehydes, acids, esters, phenols, hydroxy phenols, anhydrides, amines, nitriles and salts of organic acids or bases(Gr. I).

S_2 = Polybasic acids, hydroxy acids, glycols, polyhydroxy aldehydes, polyhydroxy ketones, sugars, amides, aminoacids, polyamino compounds, sulphonic acids, sulphinic acids, few metallic salts of organic acids and inorganic acids, salts of organic bases (Gr. II).

A_1 and A_2 = Acids, phenols, 1° and 2°-nitro compounds, oximes, mercaptans, thiophenols, sulphonic and sulphinic acids, sulphonamides, a few diketones and β-keto esters (Gr. III).

B = Amines (1°, 2° and 3°), hydrazones (Gr. IV), (Basic compounds).

I = Saturated aliphatic and alicyclic hydrocarbons, aromatic hydrocarbons and their halogen derivatives and diaryl ethers (Gr. VI). (inert compounds).

M = Nitro, nitriles, amides, derivatives of carbonyl compounds, nitroso, azo, hydrazo, sulphones, and amines substituted by negative groups (Gr. VII) (miscellaneous compounds).

N_1 and N_2 = Unsaturated hydrocarbons, polyalkylated aromatic hydrocarbons, high molecular weight alcohols, ketones, aldehydes, esters, anhydrides, ethers, acetals, lactones, acylhalides (Gr. V).

This list of compounds should match the elements present in the compounds (vide element detection tests). The class of compounds with the missing element(s) should be omitted in the LNB while making the list.

(E) *Tests for functional groups:* To be done in the following order.
(a) S = Sample, (b) "Discussions" (2.5) to be consulted before each experiment.

Sl No.	Experiment	Observation	Conclusion
o(a)	S in dry C_6H_6 + I_2 in dry C_6H_6 soln. (0.005%)	(i) Violet/deep blue (ii) Light brown shade	(i) Oxygen is absent in S. (ii) S contains oxygen.
o(b)	S + Ferrox reagent	(i) Insoluble (ii) Deep red/rose red solution	(i) Hydrocarbon or halogenated hydrocarbon. (ii) S contains O/S/N (any one or more than one).
1(a)	S/H_2O or S/EtOH + moist litmus paper.	(i) Blue litmus turns red (ii) Red litmus turns blue	(i) S is acidic. (ii) S is basic.
(b)	S + aq. $NaHCO_3$ (5% soln.)	Evolution of CO_2 as bubbles.	Strong organic acid (–COOH).
(c)	**S/H_2O or S/pure EtOH + KI/H_2O (2% soln.) + KIO_3/H_2O (4% soln.), shaken well and if required warmed, cooled and treated with freshly prepared starch soln. (Iodate-iodide test)**	Deep blue colour due to starch–iodine complex.	The sample must be acidic.
(d)	Na_2CO_3/H_2O (0.1 N, 1 ml) + 0.5% phenolphthalein (2 drops) → pink colour + S/H_2O (10 ml).	Pink colour discharged.	S is acidic.
(e)	(i) S/H_2O (soln.) + dil. HCl (X's) (ii) S/H_2O (soln.) + dil. NaOH (X's)	(i) Ppt. (ii) Ppt.	(i) S is salt of metal. (ii) S is salt of inorg. acid.
2.	If S is neutral (Test 1): S + pure EtOH or H_2O + solid $AgNO_3$ (pinch), shaken well and tested with blue litmus.	Litmus paper turns red.	Possibly S contains–C ≡ CH group.
3(a)	S + very dil. $KMnO_4$ soln. + dil. H_2SO_4 (cold or warmed).	Purple colour discharged.	S is unsaturated and/or contains easily oxidisable functional grs.
(b)	S + aq. soln. of Br_2 (v. dil.)	Yellow colour is discharged.	S is unsaturated or phenol or phenol-derivative.
(c) (i)	S/dry CCl_4 + Br_2/CCl_4 (dil).	The soln. becomes colourless,butHBr is *not* elvolved (Ag NO_3 test)	Unsaturated compound (Br_2 addition).
(ii)	S/dry $CHCl_3$ + $Br_2/CHCl_3$ (dil).	– do –	– do –
(iii)	S/dry CS_2 + Br_2/CS_2 (dil).	– do –	– do –
(iv)	S/dry CCl_4 or $CHCl_3$ or CS_2 + Br_2 in same solvent	Br_2 Colour discharged but HBr gas is evolved.	S contains replaceable H.

4(a)	S/H_2O + $FeCl_3$/H_2O (freshly prepared dil. solution).	Blue/green/purple colour developed. (i) Discharged by HOAc. (ii) Not discharged by HOAc.	 (i) S is phenol/enol. (ii) S is salicylic acid or similar compd.
(b)	S/dil. NaOH + diazotised aniline HCl soln.	Red dye	S is phenolic compd.
(c)	S/H_2O + Ferric–Ferricyanide soln.	(i) Deep blue ppt. or soln. (ii) The light brown colour of the reagent remains unaltered.	Phenols/phenolic acids/aromatic amines. S may be enol or aliphatic amine but not phenolic compd (cf. 4 (a) i).
5(a)	S/H_2O + Ceric ammonium nitrate (i.e. CAN) soln.	Colour changes to deep yellow or brown	S may be alcohol or a few phenol.
(b)	S + 3,5-Dinitrobenzoyl chloride	Solid derivative insoluble in H_2O.	Alcohol/Phenol/Amine.
(c)	S/H_2O or S/Pure MeOH + I_2 + dil. NaOH (warmed to 65–70°C, steam bath)	Yellow ppt. with characteristic smell of CHI_3	R–CH (OH) Me/R–COMe/EtOH/ MeCHO any one is the S.
6(a)	S/MeOH (pure) + Semicarbazide hydrochloride + Pyridine/Pure MeOH, warmed and poured onto cold water or S + Semicarbazide hydrochloride + acectone-free MeOH + NaOAc warmed and poured onto H_2O.	White ppt. (crystalline or amorphous) of semicarbazone (S-one).	S is carbonyl compound.
(b)	S + hydroxylamine. HCl + pyridine or NaOAc in MeOH mixture, warmed and poured onto H_2O.	White ppt. (crystalline or amorphous) of oxime	–do–
(c)	S + Pure EtOH + 2,4-dinitrophenylhydrazine HCl (or H_2SO_4) in EtOH (2,4-DN Ph. H_2SO_4). soln	(i) Yellow ppt. of 2,4-dinitrophenyl-hydrazone (2,4-DNP deriv.) (ii) Red/orange ppt. of 2,4-DNP deriv.	(i) S is *not* α,β-unsaturated carbonyl compd. S is α,β-unsaturated carbonyl compd.
7(a)	S + Schiff's reagent.	Light red colour is restored.	S is aldehyde.
(b)	S + Fehlings soln. (Mixture of Fehling A and B in equal vol.), warmed (short time).	Red ppt. (of Cu_2O),	–do–
(c)	S + Tollen's reagent (ammoniacal $AgNO_3$) warmed in steam bath (short time).	Silver mirror (if not shaken) or black Ag powder ppt. (if shaken).	–do–

8.	S/H_2O + α – naphthol ($CHCl_3$ soln, 10%, 2 drops added), to this mixture is added conc. H_2SO_4 through a dropper without shaking (Molische Test).	Violet ring at the junction of the two liquids ($CHCl_3$ and H_2SO_4).	S may be a carbohydrate.
9.	S + aq. NaOH (cold).	Clear solution, sometimes ppt. on addition of inorganic acid.	Anhydride, ester or lactone.
10.	S + $NH_2OH{\cdot}HCl$ + aq. NaOH (10 %, X's, if needed a little EtOH added), boiled (5 mins.), cooled, acidified the R.,M, with dil. H_2SO_4, then dil. $FeCl_3$ added (4–5 drops), [Hydroxamic acid test for esters].	Violet/purple/deep brown coloured soln. is obtained.	Ester/a few acid chloride/anhydride. (Phenol does not interfere).
11.	S + constant boiling HBr or HI (↑↓), neutralised and R.M. treated with dil. $FeCl_3$.	Violet/purple colour of the soln.; discharged by dil. HOAc and also white ppt. with PhCOCl in alkaline medium.	Test for phenol is positive. S *is* aromatic ether Ar–O–R.
12.	S/EtOH (dry) + Picric acid in dry EtOH (sat. soln.). or S/benzene (dry) + Picric acid in dry benzene (sat. soln.). [Picric acid should be dried on filter paper/blotting paper just before use].	Deep yellow/brown ppt. of picrate, (crystalline and unstable).	S is aromatic hydrocarbon (some phenols and aromatic amines also responds).
	If N is present, the following tests are to be done. If N is absent test nos. 13 to 16 are to be ignored.	Lassaigne's Na–fusion test done again to confirm the absence of N in the sample.	The S may contain functional groups with C/H/O as well as N (when N is present).
Sl. No.	*Experiment*	*Observation*	*Conclusion*
13(a)	S + aq. NaOH (boiled, short time).	NH_3 gas evolved (smell; red litmus turns blue; Nessler reagent; white fume with conc. HCl)	Ammonium salt, aldehyde ammonia and urethane.
(b)	S + aq. NaOH (↑↓, long time).	(i) NH_3 evolved quickly/in large amount. (ii) NH_3 evolved very slowly and in small volume.	(i) S is simple amide. (ii) S is nitrile.

14(a)	S + aq. HCl (dil. cold).	Soluble but initially it was insoluble in H_2O.	S is an amine (1°, 2° or 3°).
(b)	S + aq. HCl (dil.) + $NaNO_2$ (dil. ice–cold soln).	(i) Quick evolution of N_2 (ii) No N_2 evolution, only oily or solid ppt.	(i) 1° aliphatic amine. (ii) 2°/3° amine
(c)	The R.M. in 14(b) is taken and added to ice–cold alkaline β-naphthol solution.	Red – azodye.	1° – aromatic amine.
(d)	If 14(c) is negative, S + conc. HCl (in special case 70% H_2SO_4), boiled (5 min), cooled, diluted and 14(c) is again done [after doing the 14(b)].	Red dye	S is anilide.
(e)	If 14(d) is negative, S + Sn + conc. HCl, boiled till a clear soln. is obtained. The R.M. diluted, cooled and successively 14(b) and 14(c) done on it.	Red dye	S is aromatic nitro compound
(f)	S + aq. NaOH (strong) heated (when S is almost colourless)	The R.M. became yellow or brown	–do–
(g)	If 14(c) is postive, the R.M. of 14 (b) + EtOH + conc. HCl, boiled, cooled, diluted and again 14(c).	Again red dye	S is aromatic compound with 1°–amino as well as anilide groups. $Ar]^{-NH_2}_{-NHCOR}$
(h)	S + conc. HCl (2 drops), mixed well, then added $K_2Cr_2O_7$ (solid).	Rose-red colour	S is an anilide.
(i)	If 14(c) is positive but 14(g) is negative, the R.M. of 14(b) + EtOH (boiled) + conc. HCl + Sn (again boiled), cooled, diluted and successively repeat 14(b) and 14(c).	Red dye	S is an nitroaniline $Ar]^{-NH_2}_{-NO_2}$
(j)	If 14 (c) is positive (or negative), S + Zn (powder) + NH_4Cl/EtOH, boiled, filtered and the filtrate + Tollens reagent (heated in a steam bath). [Mulliken–Barker Test]	Silver mirror or black Ag powder ppt.	S is nitro with or without 1°–amino group on aromatic nucleus. $Ar]^{-NH_2}_{-NO_2}$ /Ar] – NO_2
(k)	S + phenol + conc. H_2SO_4. The soln. poured onto H_2O and NaOH (dil. X's) added to it.	Blue colour developed	S is nitrosoamine $R_2N{\cdot}NO$.

(l)	(i) S + aq. NaOH (r.t) + PhCOCl (shaken). (ii) S + Ac_2 O + AcONa, heated and poured onto cold H_2O. (iii) S + Ac_2 O + Pyridine (heated and cooled) and poured onto cold H_2O (iv) S dissolved in dil. HCl + Ac_2O + NaOAc (immediate addition, shaken). (Fieser's method)	(i) Solid benzoyl deriv. (ii) Solid acetyl deriv. (iii) –do– (iv) –do–	(i) 1°/2° amine (also phenols). (ii) –do– (iii) –do– (also R – OH/ArOH) may respond. (iv) 1°–aromatic amines.
(m)	(i) S + CH_3I (shaken), crystallisation from Et_2O (care! highly inflammable). (ii) S + aq. soln. $K_4Fe\ (CN)_6$. (iii) S + dil. HCl + aq. $NaNO_2$ (ice–cold), shaken well, and dil. NaOH (X's) added. The ppt., thus obtained, filtered. The residue reduced with Zn + dil. HCl then filtered. The filtrate after treatment with H_2S (passed through the soln.) was treated with $FeCl_3$ soln. 2/3 drops).	(i) Solid methoiodide (insoluble in Et_2O). (ii) Precipitate (iii) Deep blue or deep brown coloured soln. (this is thionine dye).	(i) S is 3°–amine. (ii) Maybe 3°–amine. (iii) S is aromatic 3°-amine
15.	S warmed on steam bath in a porcelain basin, treated with Cl_2 water (or dil. HNO_3), cooled then treated with NH_4OH (or NaOH) (Murexide Test).	Purple/light violet colour.	S is purines.
16 (a)	S + Potassium – Mercuric iodide soln. (Reagent) (Mayer's Test).	(i) Precipitate	S may be an alkaloid.
(b)	S + aq. soln. KI/I_2 (Wagner's Reagent).	Ppt.	S may be alkaloid.
(c)	S + aq. soln. Potassium–Bismuth iodide (Dragendorff's Reagent)	Ppt.	S may be an alkaloid.

NOTES: (i) Tests for oxime, hydrazone, hydrazide, isocyanate, isocyanide. alkyl nitrate, nitrite, azo, azoxy and nitroso compound have not been included in the aforementioned functional group tests.

(ii) Even after Lassaigne's sodium fusion test done twice, test for sulphur is negative, the 17th test is unnecessary. Similarly, if test for halogens is negative test number 18 is unnecessary.

Sl.No.	Experiment	Observation	Conclusion
17(a)	S + H_2O (boiled), tested with (i) blue litmus.	(i) Acidic soln. (litmus turns red).	(i) Sulphonic acid
	(ii) aq. $NaHCO_3$ soln.	(ii) Neutral or faintly alkaline	(ii) Alphatic sulphoxide
		(iii) S insoluble in H_2O.	(iii) Mercaptan, sulphide, disulphide.
(b)	S + H_2O + aq. $HgCl_2$ soln.	solid/oily ppt.	–do–
(c)	S + dil. NaOH soln.	Precipitate	Sulphate (salt) of org. base
(d)	S + H_2O (filtered). To the filtrate $BaCl_2$ soln.	$BaSO_4$ ppt. (insoluble in HNO_3).	–do–
18 (a)	S + H_2O (boiled), cooled + aq. $AgNO_3$ soln.	White/light yellow ppt.	S is RCOHal/Ar COHal or Halide salt of org. base.
(b)	S + H_2O (boiled) soln. tested with blue litmus.	Acidic soln. (litmus turns red).	–do–
(c)	If 18(a) is negative S + $AgNO_3$ + EtOH (boiled).	(i) Ppt. obtained	(i) Easily replaceable halogen compd.
		(ii) No ppt.	(ii) Halogen is on the aromatic ring.
(d)	S + dry EtOH + KOH (powder)	(i) Ppt. obtained	(i) Aliphatic chloride or aliphatic bromide.
		(ii) No ppt.	(ii) Iodo compound (but not aryliodide).

[F] *Preliminary suggestion:* The nature (i.e. class) of the compound is to be suggested here and to state the absence/presence of heteroatom(s), aliphatic/aromatic, functional group(s) present, and the chemical character of the compound given. Only conclusion and no argument in favour of the conclusion.

[G] *Accurate determination of b.p./m.p.* of the sample (if required the sample should be purified). If the m.p. < 150° H_2SO_4 bath, and m.p. > 150° electric bath are to be used.

[H] *Literature survey* to select apposite derivatives through which the identity of the supplied sample is confirmed (aided by the b.p/m.p. of the sample determined in [G]). These data are to be posted in box as shown below.

Probable compd.(s) (Name(s))	*Reported b.p./m.p. (lit. surv.)*	*Name of suitable deriv.*	*Observed m.p. of the deriv.*
(a) ...	(a) ...	(a) ...(i) Name –m.p.	(i) ...
(b) ...	(b) ...	(ii) –do–	(ii) ...
(c) ...	(c) ...	(b)... (i) –do–	(i) ...
etc.		– (ii) –do–	(ii) ...
		(c) – (i) –do–	(i) ...
		–(ii)–do–	(ii) ...

[I] Preparation of appropriate derivative: As suggested in [H].

(a) Different types of derivatives should be prepared. If the sample holds more than one functional group, it is advisable to prepare derivatives from each one of them. Generally, *two* derivatives should suffice but if the sample is too much simple, more than two derivatives should by tried.

(b) Brief and unambiguous methods of preparations of the derivatives should be given.

(c) During purification of the derivatives the solvent, method of drying reported m.p. and observed m.p. should be mentioned.

(d) These derivatives should be taken in a sample tube, corked, labelled properly with Roll No. of the student, sample No., name of derivative and m.p. of the derivation, then submitted. This is also an important part of the study of chemistry.

[J] *Preliminary conclusion:* Nature and name of the compound. The arguments leading to identification of the sample should be given here in detail. Spectral data (U.V., 1 R., NMR, MS, etc) to be given, if available.

[K] *Special Test(s) of the sample:* If such test(s) exist for the sample it should be performed lending additional support to the conclusion.

[L] *Final Conclusion:*

(a) Name of the supplied compound (i.e. sample).

(b) Structure of the compound (identified).

(c) Any special comment on the supplied sample.

2.4 SUMMARY OF TESTS FOR FUNCTIONAL GROUPS AND DERIVATIVE PREPARATION (Page 91):

2.5 DISCUSSION ON THE IDENTIFICATION OF ORGANIC COMPOUNDS (VIDE 2.2 A TO M):

[A] Physical Properties of the Sample

(i) Magnifying glass may be useful to determine the shape (if crystalline), as well as the nature (if amorphous) of the sample.

(ii) *The m.p.* of the solid sample may be suggestive of the molecular weight of the compound as well as the functional group(s) present in the molecule, but this is not at all dependable [e.g. CH_3–CH_3 (gas), CH_3–CH_2–OH (liquid), HO_2C–CO_2H (solid, m.p. 189.5°); again malonic acid, (m.p. 136°; mol. wt. 104), succinic acid (m.p. 185°, mol. wt. 118), glutaric acid (m.p. 97°, mol. wt. 132), so the m.p. of compound is inconclusive).

(iii) On the contrary, *the b.p.* of the compound delivers more informations regarding the chemical character of the compound. Thus in a homologous series lower b.p. corresponds to lower molecular weight. The difference in b.p. will be more pronounced between the homologues of low molecular weights but if the molecular weights of two homologues are too high the difference between the b.ps becomes less noticeable. Thus,

(a) Benzene (mol. wt. 78) bp. 80° < toluene (mol. wt. 92), b.p. 111° < *o*-xylene (m. wt. 106), b.p. 144°C.

(b) Methanol (mol. wt. 32), b.p. 64° < ethanol (mol. wt. 46), b.p. 78° < *n*-propanol (mol. wt. 60), b.p. 97°C.

(c) Moreover, spherical isomer of a compound has lower b.p. than that of the linear isomer (e.g. *tert*–butyl alcohol (b.p. 83°C) < - butyl alcohol (b.p. 100°C) < *n*-butyl alcohol (b.p. 118°), these are isomeric compounds and each one has mol. wt. 74 and OH gr.

(d) Impure compound has lower m.p. than that of pure compound as the impurity acts as the solvent. In a capillary tube the pure sample does not undergo the depression of melting point because the glass does not act as a solvent. Likewise, if the pure compound is mixed with glass powder there would be no depression of melting point. So to prepare analytically pure sample of solid compound, it is repeatedly crystallised from pure solvent until the melting point becomes constant (at least for two or more crystallisations). This property of depression of melting point with

2.4 Summary of tests for Funetional Groups and Derivative preparation

Sl. No	*Group*	*Experiment (general)*	*Experiment (special)*	*Derivatives (may be submitted)*
1.	Alkane $>C-C<$	No functional gr.	Determination of % C, H; spectral analysis (n mr, m. s. etc.)	Special reactions to prepare solid derivatives.
2. (a)	Alkene $>C=C<$	(i) Br_2 becomes colourless. (ii) $KMnO_4/H_2SO_4$ soln. becomes colourless. (iii) $C(NO_2)_4$ = TNM = Tetranitromethane in EtOH becomes red.	(iv) H_2/Pt or H_2/Pd 1 mol. H_2 absorbed. (v) Ozonolysis gives oxo compds (furnish 2,4-DNP, S-one, etc).	(i) Bromo/lodo comp. (if solid) (ii) $KMnO_4$/NaOH soln. oxidised deriv. if solid. (iii) S-one/2,4-DNP from ozonolysis.
2. (b)	Alkyne $-C\equiv C-$ and $-C\equiv C-H$	(i) Vide 2(a) (ii) Tollens reagent (iii) Cu_2Cl_2/NH_4OH reagent.	(iv) H_2/Pt or H_2/Pd 2 mols H_2 absorbed.	–do–
3. (a)	Alkanol R–OH	(i) $KMnO_4/H_2SO_4$ soln. becomes colourless. (ii) Gives fragrant esters with $R-CO_2H$/conc. H_2SO_4.	CAN = Ceric ammonium nitrate, in aq. soln. turns red or brown.	(i) Acetate (ii) Oxidised and 2,4-DNP/S-one of the oxo product.
3. (b)	EtOH and R – CH (OH) – CH_3	(i) EtOH only 1°–alcohol giving CHI_3 (ii) Methyl carbinols giving CHI_3 (iii) Sweet smelling esters from EtOH	(i) CAN Test positive $(NH_4)_2[Ce(NO_3)_6]$ (ii) $KMnO_4/H_2SO_4$ dil. becomes colourless (warming fast). (iii) Cr^{VI}/H_2SO_4 turns green (warming).	(i) CHI_3, yellow, mp. 119°C. (ii) 2,4-DNP/S-one *via* oxidation
3. (c)	3(a) and 3(b)	Xanthate (sample, 100 mg + solid KOH, 500 mg + CS_2 few drops + $(NH_4)_2MoO_4$, trace, shaken. The R.M. acidified with dil. HCl and treated with $CHCl_3$, 1ml. Organic layer turns violet).		

4.	Alkanal (i) CH_3-CHO and $R-CHO$ (oxo)	(i) Only aldehyde giving CHI_3. (ii) $KMnO_4/H_2SO_4$ soln. colourless. (iii) Tollens Test postive. (iv) Br_2 soln. colourless (v) Fehling's soln. gives red Cu_2O.	Responds to Schiff's Test and other oxo-reagents.	(i) CHI_3 from CH_3CHO. (ii) 2,4-DNP/S-one/oxime. (iii) $R-COOH$ if solid (via oxidation)
5.	Alkanone $C-\overset{\overset{O}{\parallel}}{C}-C$ (oxo)	(i) Br_2 soln. colourless. (ii) $KMnO_4/H_2SO_4$ soln. colourless (warm.)	(i) Legal's test positive. (ii) $C-\overset{\overset{O}{\parallel}}{C}-CH_3$ gives iodoform.	(i) 2,4-DNP/S-one/oxime, etc. (ii) CHI_3 from methylketones. (iii) $R-CO_2H$ (via oxdn.) if solid.
6.	Ether C–O–C (oxa)	Const. boiling HI generates RI and R′–OH or Ar–OH.	(i) Alkyl iodide and alcohol/phenols identified if possible. (ii) Ferrox reagent test (if sl. no. 3 to 5 are negative).	(i) From alcohol via oxdln. 2,4-DNP or S-one. (ii) From phenol benzoate derivation.
7.	Phenol Ar–OH	(i) $FeCl_3$ soln. Violet/brown/blue colour. (ii) Br_2 water colourless. (iii) Azodye test positive. (iv) Ferric–Ferricyanide test positive. (very sensitive and trustworthy). (v) Liebermann test positive.	1. Phenolphthalein indicator preparation with dil. alkali (Phenol → pink), (o – cresol → red), (m – cresol → blue – red), (resorcinol → green), (catechol → blue), (hydroquinone → deep violet), (α-naphthol → deep green), (α-naphthol ↔ light green).	(i) Benzoate (Schothen–Baumann) (ii) Acetate (iii) Bromo derivatives
8.	Alkanoic acid R–COOH	(i) $NaHCO_3$ soln. gives CO_2 ↑ (ii) Blue litmus turns red. (iii) Congo-red paper (no change).	(i) KI + KIO_3 soln. + Starch turns deep violet (very sensitive). (ii) Pink colour of phenolphthalein + Na_2CO_3 soln. becomes colourless. (iii) Sweet smelling esters with ROH/conc. H_2SO_4 (warming)	(i) Acid amide $R-CO-NH_2$, (ii) Anilide, R–CO–NH–Ar, $Ar-CO-NH_2$, $Ar-\underset{\underset{O}{\parallel}}{C}-NH-Ar$.

9.	Alkanoate or Esters $-C(=O)-OR$	(i) Sweet smell when low mol. wt. (ii) Saponification and isolation of organic acid (if solid) and identification of alcohol part via oxidation to oxo.	(i) Hydroxamic acid test (positive) (ii) With phenolphthalein + (0.1 N) NaOH (1 drop) heated, becomes colourless.	(i) If EtOH obtained after saponification CHI_3 is obtained. (ii) Acid part after saponification. (iii) 2,4-DNP/S-one via oxo from R–OH.
10.	Aromatic hydrocarbons. Ar–H.	(i) I_2 in C_6H_6 test. Violet. (ii) Ferrox test (negative)	(iii) Oxidation to deliver quinones (iv) Nitration to deliver nitro-compounds	(i) Nitro if solid (ii) Quinones. (iii) Picrates.
11.	Amines 1°–, 2°, and 3°	Hinsberg Test:– Sample (100 mg) + dil. NaOH (5%, 5 ml) + Pyridine (2 ml), shaken and benzene sulphonyl chloride, C_6H_5–SO–Cl (few drops) added and shaken well and the following colour changes are noted:		
		Amine	1°-($-NH_2$)	2°-(>NH)	3°-(–N–)
		Colour after Hinsberg Test	yellow	orange	Red/purple-blue –red
11. (a)	1°-Amine $Ar-NH_2$	*Experiment* (i) Carlylamie Test (positive) (ii) Azo dye test (positive). (iii) Soluble in dil. HCl. (iv) Dil. HNO_2 (diazotised, N_2-↑ from aliphatic amine)	*Experiment* (i) 2,4-DNF B Test. (on filter paper) Sample (1 drop) + 2,4-DNF B (sat. soln. 1 drop), turns deep yellow $-NH_2 + F-C_6H_3(NO_2)_2$ $\rightarrow -NH-C_6H_3(NO_2)_2$ (deep yellow)	*Derivatives* (i) Acetyl. (ii) Benzoyl, (ii) Bromo.

11 (b)	2°-Amine. >NH	(i) Dil. HCl (soluble). (ii) Dil. HNO_2 (light yellow >N.NO).	(i) Liebermen Test (positive) (ii) Simon's Test (positive).	(i) Acetyl. (ii) Benzoyl.
11 (c)	3°-Amine. –N– |	(i) Dil. HCl. (soluble) (ii) Dil. HNO_2 (deep yellow ppt.)	(i) Potassium ferricyanide test (yellow ppt)	(i) Picrate (ii) Metho-iodide (solid)
12.	Amino acids. Q] (NH_2) (CO_2H).	(i) Tests for both NH_2 and CO_2H are positive.	(i) Ninhydrin Test (colour) 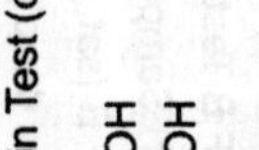 Various types of amino acids give different colour.	1. Derivatives of the $-NH_2$ group.
13.	Amide Q–CO–NH_2 (Q = R/Ar)	(i) Dil. HNO_2 (N_2↑) (ii) Neutral to litmus. (iii) Dil. NaOH, Δ.(NH_3-↑)	(i) 1° Amine *via* Hofmann degradation and azo dye text Carbylamine test.	(i) Organic acid *via* hydrolysis. (ii) Deriv. of 1° amine *via* Hofmann degradation.
14.	Anilide Ar–NH–COQ	(i) $ArNH_2$ *via* hydrolysis, then azo dye and carbylamine.	(i) Tafel's Test: Sample + conc. H_2SO_4 + $K_2Cr_2O_7$ (blue or purple colour).	(i) R–COOH *via* hydrolysis. (ii) Acetyl and benzoyl of Ar–NH_2
15.	Nitrile Q–CN (Q = R/Ar)	(i) Neutral (ii) Conc. NaOH, Δ (NH_3-↑)	(i) Reduction to R–CH_2–NH_2 and experiment	(i) Q–COOH *via* hydrolysis. (ii) Acid amide *via* H_2O_2 oxidation.
16.	Nitro Ar–NO_2	(i) Neutral (ii) Sn + conc. HCl, Δ (Ar–NH_2 → azo dye and carbylamine)	(i) Reduction and test for 1° – amine (ii) Mulliken–Barker Test.	(i) Dinitro/Trinitro. (ii) Deriv. of 1°-amine *via* reduction

impurities has been utilised in establishing the identity of a compound by mixing the compound with the authentic pure sample (preferably in 1:1 ratio) and if there is no depression of the mixed m.p. (m.m.p) then it would be concluded that the sample is identical with the authentic sample (but it suffers from the limitation that the authentic sample might not be always available). An example is cited here to clarify the point. A compound melts at 120°C. This is mixed with β-naphthol (m.p. 122°C) and the m.m.p. is depressed, i.e. the mixture melts at much below 120°C. But when the compound (m.p. 120°C) is mixed with benzoic acid (m.p. 122°C), there is no depression of m.m.p. then it will be concluded that the unknown compound must be benzoic acid.

(iv) *The boiling point* of a compound is elevated by impurities. Therefore, by repeated distillation the constant boiling sample is considered to be pure, and such sample is analysed. It should be understood that the b.p. of an impure liquid is spread over a range of temperature, while the b.p. of pure liquid is very sharp and varies only in 2 to 3°C (taken to be constant boiling).

(v) *Smell or odour* of the sample may also help its identification (guess but not confirm). The odour is expressed as "characteristic", sweet, fragrant, pleasant, unpleasant, agreeable, disagreeable, sharp, pungent, nauseating, etc. Thus, ethanol has its characteristic odour, ethyl acetoacetate has the odour of ripe jujube and methyl saifcylate has the odour of oil of wintergreen (iodex). Generally esters have somewhat fruity-sweet odour, ketones are fragrant. N-containing compounds have unpleasant odour and S-compounds have disagreeable odour.

To determine the relationship between odour and chemical constitution various compounds with mol. wt. ranging between 17 and 300 have been examined. A few functional groups imparts odour to the compound. These functional groups are called *odoriphores* (of i.e.. cromophores, brings colour). Odoriphores which bring sweet smell to the compounds are called *eusomophores* (e.g.–OH,–OR.–CNO,–CO_2R,–CN etc), on the contrary, the functional groups which bring unpleasant smell to the compounds are called *cacosomophores* (e.g. –SH, SR, COSR, NC, etc.). The higher homologues are more odourless compared with their lower homologues, but the higher homologues if made unsaturated by introducing olefinic bond become again odoriferous. The smell also depends on the size of the alicyclic rings, e.g. if the cycle is 15-membered, the musk-odour will appear, e.g. (A), (B) and (C).

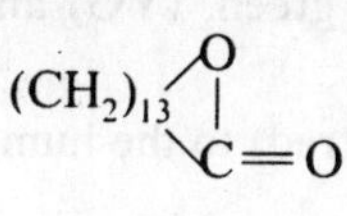

(A)

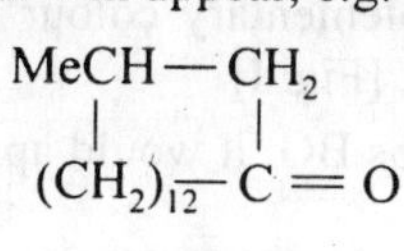

(B)

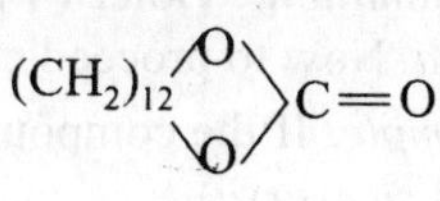

(C)

Again it has been observed that although –CN, –CHO and NO_2 groups are entirely different yet they impart almost similar smell to the molecules. The smell of a compound is also dependent on the volatility, power of adherence to the fats, and the internal vibrations of the molecules. Spherical molecules having diameter 7 Å smell like camphor. The primary odours are camphoraceous, pungent, ethereal, floral, piperminty, musky and putrid which are often mentioned.

(vi) *The colour* of the compound also suggests some chemical constitution, unsaturated or unstable compound easily becomes dark (brown or black) due decomposition under the influence of light, oxygen or long standing at room temperature. On distillation of the liquid sample or crystallisation of the solid sample, the original colour of the compound reappears. The stable coloured compounds contain $-NO_2$, $-NO$, $-N=N-$, quinonoid group and $>C=C<$ or $-C \equiv C-$ groups which are termed *chromophore* groups imparting colour to the compound. In addition to chromophore groups the molecule can contain –OH, $-NH_2$, –Cl, $-CO_2R$ etc., which are called *auxochrome* groups which deepen the colour brought by the chromophore groups. Those groups which increase the wave length in the absorption spectra of the compounds are called *bathochromic* group and those groups which reduce the wave-length are called hypsochromic groups.

Human eye can detect the colour of the compound which can absorb light of wave length between 400–800 nm. If compound cannot absorb any wave length of this range, this will appear as colourless while if a particular wave length (or a small range) is absorbed by the compound, the *complementary colour* of the absorbed light will be perceived by the observer. Again if the whole range (400–800 nm) of wave-length is absorbed by the compound it will appear as black. Thus,

Absorbed light by the compound. wave length	*colour absorbed*	*Colour seen by the observer (Fig. I)*
(a) 400 nm ...	(a) Blue ...	(a) Yellow
(b) 500 nm ...	(b) Green ...	(b) Red
(c) 600 nm ...	(c) Yellow ...	(c) Blue
(d) 700 nm ...	(d) Red ...	(d) Green

The following mnemonics would be helpful to remember the order of the colour with the increasing absorbed wave length of light.

(a) V– IB – BG – G – YG – Y – O – R – P (*vib. bg g yg yorp*).

(b) V (violet, 400 nm), I B (indigo-blue 435 nm), BG (blue-green, 480 nm), G (green, 500 nm), YG (yellow-green, 560 nm), Y (yellow, 580 nm), O (orange, 600 nm), R (red, 725 ± 25 nm), P (Purple).

(c) Complementary colour (mnemonic – very easy = *v*. Easy = *v. yg*, start)

Explanation: Violet (V) is complementary colour of yellow green, (YG) and *vice versa*. Now to proceed clockwise. [Fig. I]

Example: If the compound absorbs BG, it would appear as R (red) to the human eye (and *viceversa*).

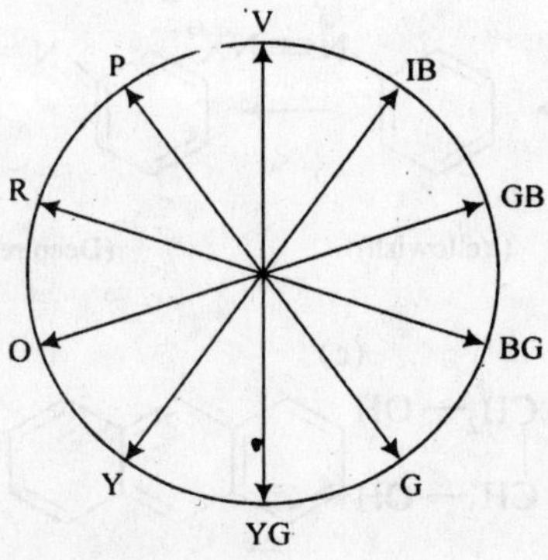

[I]

(d)

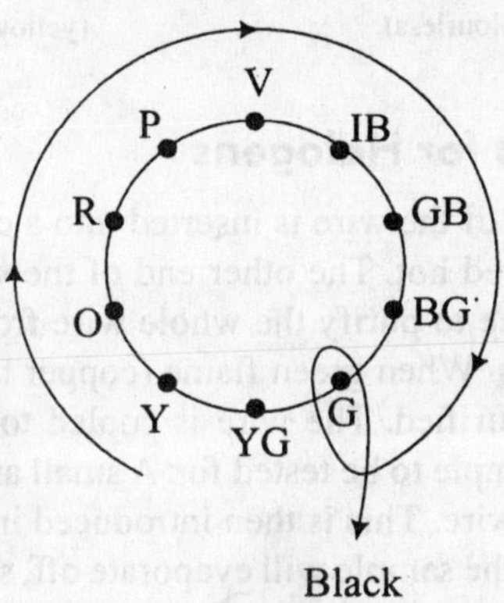

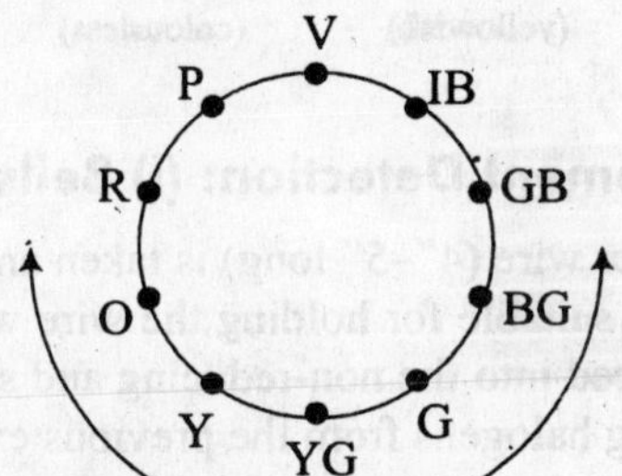

The colour is deepening in this route, i.e., YG → Y → O → R → P → V → BG → G → black (deepening of colour taking place).

Colour shift. Bathochromic in this direction, i.e. Y to R is bathochromic shift (red shift).

Colour shift Hypsochromic in this direction, i.e. YG to BG is hypsochromic shift (blue shift).

Explanation: If the compound absorbs violet (V, 400 nm, wave length) then the visible cololour of the compound will be yellowish green (YG), as shown by the double headed arrow inside the circle (Fig. I) connecting V and YG. Similarly, if the compound absorbs wave length 600 nm, this is absorbing orange (O) colour so the complementary colour bluish green (BG) will be visible to the human eye. This is shown by the double-headed arrow connecting O and GB. It may be concluded that O and GB are complementary to each other and if the complementary colours are mixed together this will result in white light (colourless).

It has been observed that when the molecule holds several conjugated double bonds or more than one oxo (i.e. C = O) group the first appearance of colour starts with pale yellow. Thus in $R(CH = CH)n - R_1$, when $n < 8$, colour UV., $n = 8$, colour YG; $n = 9$, colour Y; $n = 10$, colour O; $n = 11$, colour R (deepening of colour with the increase of number of conjugated double bonds). The following examples are also *illustrative:*

(a) 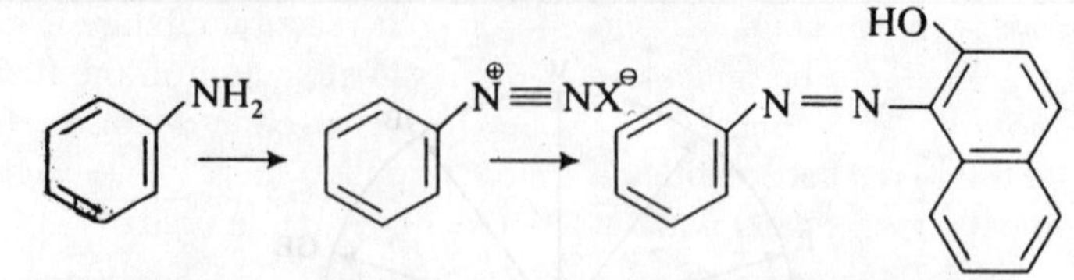

(Almost colourless) (Yellowish) (Deep red, azo dye).

(b) 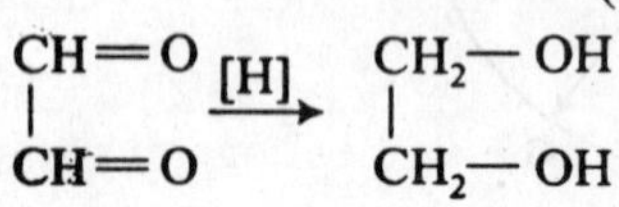

(c) 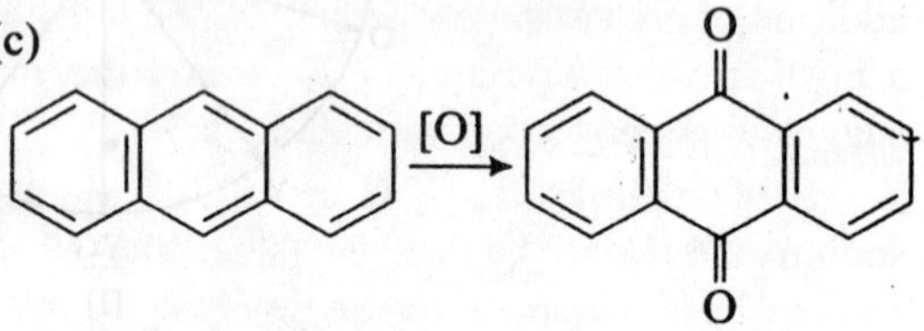

(yellowish). (colourless) (Almost colourless) (yellow)

[B] Element Detection: (i) Beilstein Test for Halogens

A copper wire (4″–5″ long) is taken and one end of the wire is inserted into a cork to make it suitable for holding the wire when it is red hot. The other end of the wire is introduced into the non-reducing and strong flame to purify the whole wire from the adhering halogens from the previous experiments. When green flame (copper halide) is observed no more the wire is granted to be purified. The wire is cooled to room temperature and the free end is dipped into the sample to be tested for. A small amount of the sample (liquid or solid) will adhere to the wire. This is then introduced into the non-luminous flame. If the wire remains too hot, the sample will evaporate off, so cool the wire. Again in some cases the solid sample may not adhere to the cold wire, here the wire is to be warmed and put on the sample which may melt and stick to the warm (not hot) wire. It is desirable that about 1 cm of the wire holds the sample.

Now the end of the wire coated with the sample is introduced in the flame when the sample will burn making the copper halide which will evaporate with a green flame. So the formation of the volatile copper salt generates the green flame. Therefore, any compound (eg. urea) which can produce volatile copper salt will give the green flame. Therefore, production of green flame in the Beilsteine Test is not a guarantee for the presence of halogens in the sample. But the absence of the green flame in this experiment confirms the absence of halogen. Thus, it is concluded that if there is green flame, the halogens may be present, but when the green flame is absent the halogens must be absent.

(ii) *Lassaigne's Sodium Fusion Test* (preferably done after Beilsteim Test): Metallic sodium is kept under kerosene as a precautionary measure. A pea-sized sodium is dried on a blotting paper or a filter paper and taken in a dry fusion tube followed by sample (ca 150 mg, if solid, and if liquid 3–4 drops) and heated slowly in a small flame initially. Then the fusion tube is strongly heated until red hot and quickly immersed under cold water (15 ml) contained in a mortar. The reaction mixture is ground well and filtered (I). It should be noted that if the residual sodium in the fusion tube is dipped into cold water, there may be an explosive reaction when the fusion tube becomes cold. But in red hot condition the molten sodium (m.p. 97°C) reacts with cold water less violently (practically without any risk of getting injured).

There is a suggestion of melting the sodium (m.p. 97°C) in the fusion tube then the sample is added to the molten sodium. But it has been observed frequently that on addition of cold sample (specially when it is liquid) to the molten sodium contained in a hot fusion tube it breaks down and the sample catches fire. So this alternative suggestion is not recommended here.

If the sample has $-OH$ or $-CO_2H$ group, there will be vigorous reaction with sodium and H_2 will be evolved. When the $-NO_2$ group is present, the sample may catch fire and burn inside the fusion tube and the heat generated may crack down the fusion tube. In such case the fire should be immediately blown out without fear. The fusion tube is to be heated slowly and on a small flame to prevent the escape of the sample as vapour before decomposition.

It is to be noted also that if the sample contains both N and S, then the heating for a prolonged time may lead to the formation of sulphocyanide (NaSCN) which will interfere with the detection of N and /or S (giving red colour with $FeCl_3$ soln.). So the fusion tube is immersed into water as soon as it becomes red hot.

The aforementioned filtrate (ca 15 ml, I) is collected in a clean test tube (t.t.) and the following tests are carried out taking 1–2 ml for each experiment.

(a) to the filtrate (1 ml) is added freshly prepared soln. of sodium nitroprusside (a small crystal of the reagent is dissolved in water just before use). Formation of red–purple coloured solution indicates sulplur in the molecule

The filtrate (1 ml) is treated with $FeCl_3$ soln. Red soln. indicates NaSCN.

(b) To the filtrate (1 ml) is added excess of HOAc (dil.) to acidify the filtrate (litmus test) followed by a solution of lead acetate. Formation of black solid lead sulphide indicates the presence of sulphur in the compound.

(c) If sulphur is detected in the filtrate then the following procedure should be helpful to detect the N in the filtrate (present as NaCN).

(a) In a t.t. $FeSO_4$ (0.2–0.3 g) is taken and washed with H_2O. The residual purified $FeSO_4$ is dissolved in 2–3 ml of H_2O to furnish freshly prepared $FeSO_4$ soln. which is added to the sodium fusion filtrate (I, 2–3 ml). Black ppt. of Fe_2S_3 is formed which is filtered, and the filtrate is again treated with $FeSO_4$ soln. to determine the complete removal of S [present as Na_2S in the filtrate (I)]. The reaction mixture is boiled (1–2 mins.), cooled (under tap water), treated with $FeCl_3$ soln. (1 drop) and acidified with dil. H_2SO_4 when blue ppt. of "Prussian Blue" is formed which soon settles down. Dil. H_2SO_4 is better than dil. HCl because $FeCl_3$ is yellowish solution in which the blue ppt. is less, pronounced, particulary if the N content in the sample is low. In such case, it is desirable to filter the solution when solid Prussian blue sticks to the white background of filter paper thus making the detection more comfortable.

(d) In a t.t. equal volumes of dil. HNO_3 and dil. $AgNO_3$ are mixed. If this mixture remains transparent and colourless then it will be suitable for the detection of halogens in the filtrate (I) because this demonstrates the absence of impurities, particularly the halides, in water and HNO_3.

To this acidified $AgNO_3$ solution is added the filtrate (I, 1 ml) and the formation of a ppt. (insoluble in HNO_3 as this is already present with $AgNO_3$) indicates the presence of halogen(s) in the sample. If the ppt. is curdy white and completely soluble in excess of NH_4OH, but reappears on acidification then Cl is present in the sample. It the halide ppt. (AgHal) is pale yellow and sparingly soluble in NH_4OH (dil.) then the halogen is Br, and if the ppt. is yellow and completely insoluble in NH_4OH (dil.) then the halogen is iodine. When a single halide is present and N is absent in the sample, this procedure may work to some extent.

(e) If the sample contains N then in the filtrate the halogen(s) is tested for after the removal of N as volatile HCN in the following way. To the filtrate (1 ml) is added dil. HNO_3 (pure) to acidify the mixture which is boiled for 2–3 mins, then cooled and dil. $AgNO_3$ is added to it. Formation of acid–insoluble white to yellow ppt. indicates the presence of halogen(s) is the sample (AgCl—white, AgBr—pale yellow and AgI—yellow).

But there is some uncertainty in this procedure. There is no guarantee that all HCN has been removed on heating (duration of heating, quantity of HCN present, concentration of HNO_3 are all governing factors). If relatively concentrated HNO_3 is used or the reaction mixture is heated too long the HX and HNO_3 may react together thus X escaping the detection. So, this method is not very convincing.

(f) In a t.t. the filtrate (I) is taken (1 ml) and treated with formalin (1 ml) and boiled (1 min) when all HCN' is removed as volatile $CH_2(OH)(CN)$ leaving NaX (sodium halide from sodium fusion test) in the reaction mixture. This is tested with the previously prepared mixture of dil. HNO_3 and $AgNO_3$ for the presence of hologens as described before. This test will get support from the Beilstein test done before.

(g) In a separate t.t. the filtrate (I) was taken (1 ml) and treated with CCl_4 (1 ml) followed by dil. HNO_3 to make the reaction mixture acidic (litmus). Freshly prepared chlorine water was added to it dropwise with shaking. Bromine or iodine (if present) will make the CCl_4 layer light brown or yellow and purple, respectively. When excess of Cl_2 – water is added and shaken the iodine colour will disappear but bromine colour will stay. In this experiment, the Br and I can be identified. If both Br and I are absent but the filtrate responds to hologan then this halogen must be Cl (AgF is soluble in water).

(h) The following method is very suitable for detection of the halogens in the filtrate (I) although at the first glance it appears to be difficult to perform.

In a t.t. the filtrate (I-2 ml) is very carefully evaporated to dryness. To the residue (almost invisible) is added a pinch of very pure MnO_2 followed by conc. H_2SO_4 (1 ml, delivered at the bottom of the t.t. by a dropper) and a cork holding a glass-jet is fitted at the mouth of the t.t. On heating the t.t. halogens will come out through the fine jet as vapour which is passed successively on (a) moist starch paper (I_2 detected by purple colour development), (b) moist fluoresceine paper (Br_2 detected by transformation of the yellow test paper into orange or red colour) and if both (a) and (b) are negative then (c) moist starch–iodide paper (Cl_2 will liberate I_2 and the paper will turn purple due to formation of starch–iodine complex). This is presented in the following box.

Experiment	*Reagents (moist)*	*I*	*Br*	*Cl*	*Comment*
From filtrate NaX (dry)	(1) Starch paper	+	–	–	+, present
+ MnO_2 (pure) +	(2) Flouresceine paper	–	+	–	(response)
conc. H_2SO_4, Δ,	(3) Starch–iodide paper	+	+	+	–, absent
gas passed on					(no response)

(i) Equations related to sodium fusion test.

(I) Sample (C/S/N/Hal) + Na (X's) $\xrightarrow[\text{2) Water}]{\text{1) Fusion}}$ $Na_2S/NaCN/NaX + NaOH$ + C → Filtrate (I)

(II) *For sulphur:* (i) $\underset{\text{(Filtrate)}}{Na_2S} + \underset{\text{(Sod. nitroprusside)}}{Na_2[Fe(CN)_5(NO)]} = \underset{\text{(Pureple colour)}}{Na_4[Fe(CN)_5(NOS)]}$

(ii) $Na_2S + \underset{\text{(Lead acetate)}}{Pb(OCOCH_3)_2} = \underset{\text{(Sod. acetate)}}{2\,NaOCOCH_3} + \underset{\text{(Black ppt.)}}{Pb\,S\downarrow}$

(III) *For nitrogen:* $\underset{\text{(NaCN)}}{2NaCN} + \underset{\text{(Filtrate)}}{FeSO_4} = Na_2SO_4 + \underset{\text{(Ferrous cyanide)}}{Fe(CN)_2}$

$$Fe(CN)_2 + 4\,NaCN = \underset{\text{(Sod. ferrocyanide soln; colourless)}}{Na_4[Fe(CN)_6]}$$

$$3Na_4[Fe(CN)_6] + 4FeCl_3 = \underset{\text{(Prussian blue; ppt.)}}{Fe_4[Fe(CN)_6]_3} + 12\,NaCl$$

(IV) *For S and N:* $\underset{\text{(giving NaSCN on strong heating)}}{3NaSCN} + FeCl_3 = 3NaCl + \underset{\text{(Purple or red colour soln.)}}{Fe(SCN)_3}$

(V) *For Halogens;* (NaX) $\xrightarrow[+AgNO_3,\ \text{Pure } H_2O]{\text{(Pure } HNO_3)}$ AgX $\xrightarrow{NH_4OH \text{ dil}}$ Soluble (AgCl, white)

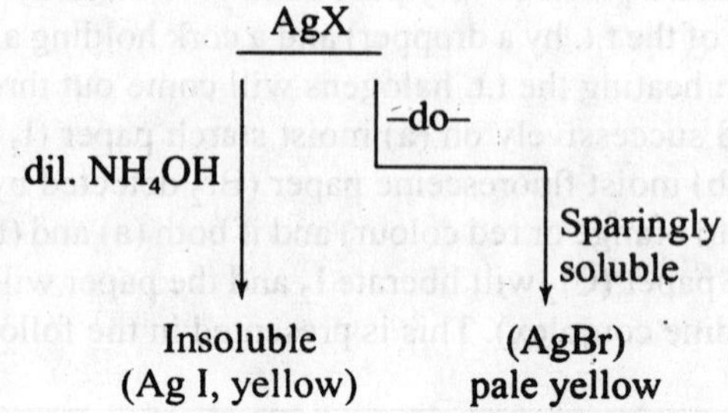

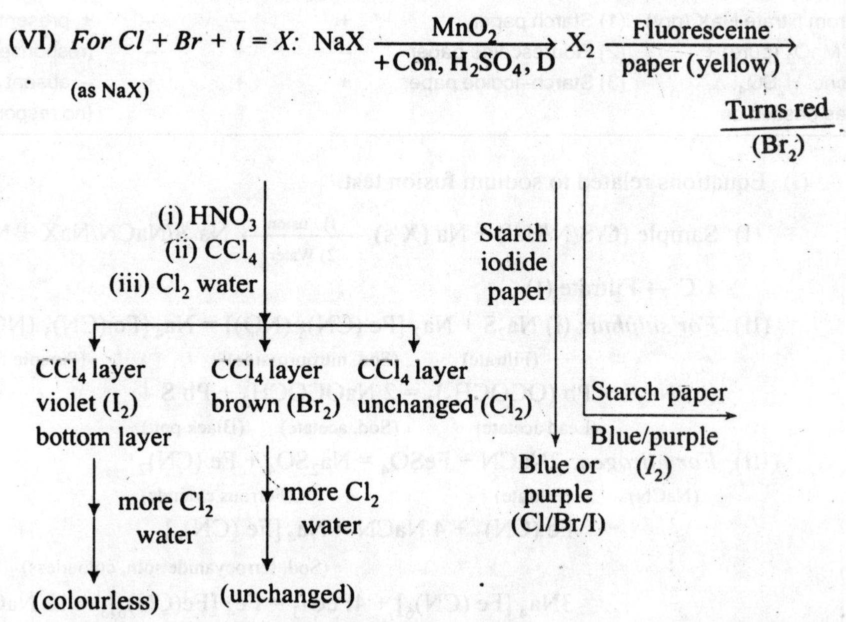

(j) Generally, it is enough to show the heteroatoms (present or absent) in the box form, shown in 2.3 [B](a). But if required to present the details of this investigation, the following form (box) may be adopted. [Summary, BB].

In a fusion tube the sample and a pea-sized sodium was taken and heated strongly and the red hot fusion tube was dipped under water contained in a mortar. The contents were crushed well and the reaction mixture was filtered. The filtrate was divided into several parts and examined.

(k) It has been suggested to follow the Zn and Na_2CO_3 method of identification of elements instead of sodium fusion method because sodium may pollute the environment. In some cases, the Zn and Na_2CO_3 fusion is really more suitable than Lassaigne's sodium fusion test. In the following the method is described.

A.R. Na_2CO_3 (25 g) is intimately mixed with powdered Zn (pure, 50 g). This mixture is ground well and a small portion is tested for S and halogens. If these

2.5 BB

Sl. No.	Experiment*(each time 1 ml filtrate) to the soln.	Observation	Conclusion (in the sample)
(i)	Freshly prepared sod. nitroprusside soln. added.	(i) Colour unchanged (ii) Purple colour	(i) S—absent. (ii) S—present.
(ii)	Dil. HOAc was added; to the acidic soln. $Pb(OAc)_2$ was added.	Black ppt. of PbS.	S—present.
(iii)	Dil. $FeCl_3$ was added	Purple or red colour was formed (soln.).	S and N both are present.
(iv)	Freshly prepd. $FeSO_4$ soln. was added, boiled for 1 min., cooled, $FeCl_3$ added (1 drop) and the R.M. was acidified with dil. H_2SO_4.	Solid ppt. of Prussian Blue was obtained. This could be filtered.	N—Present
(v)	Soln. of pure HNO_3 in pure water was mixed with dil. soln. of $AgNO_3$ and this mixture was added to one part (1 ml) of the filtrate until acidic.	(i) White thick ppt..soluble in NH_4OH (ii) Yellowish or yellow ppt. sparingly soluble or insoluble in NH_4OH.	i) N and/or Cl present ii) Br and/or I present
(vi)	Dil. HNO_3 (6N) was added until the filtrate (1 ml) was acidic, boiled (2–3 mins), cooled and dil. $AgNO_3$ was added.	Heavy white or yellow ppt. was obtained.	Probably Cl, Br, I (one or more) was present
(vii)	Formalin (5–6 drops) was added, boiled (1 min), cooled, and acidified with a mixt. of pure HNO_3 and dil. $AgNO_3$.	Ppt. was formed which was (a) white and soluble in NH_4OH, (b) yellow, insoluble in NH_4OH, (c) Yellowish, sparingly soluble in NH_4OH	(a) Cl—present (b) I—present (c) Br—present
(viii)	Filtrate (2 ml) evaporated to dryness and treated with pure MnO_2 followed by conc. H_2SO_4, heated and the issuing gas passed over moist (a) Fluoresceine paper, next (b) Starch–iodide paper and finally (c) Starch paper.	(a) Turns red or orange (b) Turns blue (c) Turns blue	(a) Br—present (b) Cl Br or I—present (all or single) (c) I—present
(ix)	CCl_4 (1 ml) added, acidified with dil. HNO_3, then Cl_2-water added.	(a) CCl_4 layer yellow (b) CCl_4 layer purple but became colourless with X's of Cl_2-water.	(a) Br—present (b) I—present

elements are absent in the Zn–Na_2CO_3 mixture, then this is suitable for use as the reagent.

The sample (0.1 g) is mixed with the Zn–Na_2CO_3 reagent in a fusion tube and this layer is covered with another portion of the reagent. The fusion tube is initially heated on a small flame, then heated strongly to red hot in the inclined position of the fusion tube. This red hot fusion tube is plunged into water contained in a mortar, ground well and filtered. The filtrate is tested for N and halogens while the solid residue retains the S as ZnS. All these elements are identified by the classical methods described before under Lassaigne's sodium fusion test [2.5 (B) ii].

[C] *Ignition Test:* A small amount (0.5 – 1 g) of the organic sample is taken on a nickel spoon and slowly introduced into the flame of the burner when the sample catches fire. If it burns with sooty flame then it will be concluded that the sample is unsaturated or it has high percentage of carbon. Most of the aromatic compounds burn with liberation of enough carbon particles but aliphatic compounds (e. g. mesityl oxide, $Me_2C = CH–CO–CH_3$) may burn like aromatic compounds. It should be noted that if large amount of the sample is taken for ignition test, the compound may liberate excessive amount of soot and carbon particles while burning, thus polluting the environment of the laboratory. In such occasion the flame should be immediately extinguished by immersing the spoon under water.

[D] *Solubility Test and Classification of the Compound*

(a) (i) The solubility of the sample should be tested with fresh sample each time at room temperature.

(ii) If the sample is soluble in H_2O (at r.t.) then testing the solubility of the sample in other reagents in aqueous solution is unnecessary. In such case in the Box [2.3 (D) (a)] the word "unnecessary" should be posted at the appropriate position.

(iii) If the sample is soluble in NaOH or $NaHCO_3$, then excess of inorganic acid should be added to the alkaline solution to check if the sample reappears or not. Likewise if the sample is soluble in dilute mineral acid then to this solution dil. NaOH solution is to be added in excess to examine whether the sample reappears or not.

(iv) If the sample is soluble in conc. H_2SO_4 (at r.t.), the solution should be poured onto cold water. If the sample reappears then it is a hydrocarbon, but if it does not reappear then the sample is an ether (mostly).

(b) If in the Element Detection Test any heteroatom is found to be absent then in the Solubility and Classification Test the letter representing the "class" of the compound will include only those which do not contain the missing element while furnishing the list of compounds belonging to that letter. Thus, if the sample does not contain N and it belongs to S_1 class, then while giving the list of compounds belonging to S_1, the amines and nitriles are to be omitted.

[E] *Test for Functional Groups*

(a) Always clean apparatus and fresh chemicals should be used. A few droppers should be kept clean and ready to use for adding dropwise of liquid reagents. Wastage of anything is undesirable. Stoppers of the reagent bottle should not be left on the working bench to avoid contact of the reagent (sometimes very corrosive) with the wooden bench. Knowledge of the poisonous and hazardous chemicals is desirable.

(b) The tests for functional groups should be done in a systematic order as shown in Sl. No. It should never be done in a haphazard way. It is important to show the presence of a particular group, but to show the absence of other groups is equally important. "Without haste and without rest" should be the motto.

(c) Whenever the appearance or disappearance of colour is indicated or uncertain ppt. appears it is always necessary to perform the blank test with only the reagents. As for example in $KMnO_4$ test for detection of oxidisable groups in the sample (Baeyer's test), a very dil. aq. sqln. of $KMnO_4$ is taken in a test tube (faint purple colour) and few drops of dil. H_2SO_4 are added to it. Half of this mixture is taken in another clean test tube to which a small amount of the sample is added. Both of these test tubes are held by the same hand, shaken well, and if not decolourised by the sample, both of these test tubes are slowly heated on a small flame with shaking and tested for decolourisation of $KMnO_4$ in the contents of the test tube with the sample. If both of them become colourless then this experiment is inconclusive.

SL. NO.

O (a) If the sample is liquid and does not contain S, N and/or halogen(s), then few drops of the sample is taken in a *dry* t.t. and *dry* benzene (1 ml) is added to it. To this mixture a solution of iodine in *dry* benzene (violet or purple colour, 0.005%, 1 ml) is added and shaken well. If the sample contains oxygen, then the solution will turn light brown or deep brown. But if oxygen is absent, there will be no change of colour of the test solution. Before performing this test on unknown sample it should be practised with ethanol, anisoles, etc. to see the colour change. This test is of minor importance, because it is very difficult to perform it in anhydrous condition.

O (b) In a mortar a few crystals of ferric alum and small amount of potassium thiocyanate are taken and ground well to make a red–brown paste. This is Ferrox Reagent.

In a clean and dry t.t. Ferrox Reagent (very small amount) is delivered at the bottom of the t.t. by means of a glass rod taking care not be touch the side of the t.t. To this is added the liquid sample (if solid, no need of this test) and shaken well. If the Ferrox Reagent does not dissolve in the liquid sample then the sample does not contain S, O or N. The sample may be a hydrocarbon or a halo-compound in this case.

On the contrary, if the Ferrox Reagent dissolves in the liquid sample (0.5 ml is enough) forming dark red solution then the sample may contain S, O and N (any one or more than one). Again if Lassaigne's sodium fusion test shows the absence of S and N in the sample yet the Ferrox Reagent dissolves and furnish red solution then the sample contains O (e.g. aromatic or aliphatic ethers, ketones, alcohols, esters, etc.).

In many cases the Ferrox Reagent is obtained as sticky mass or heavy liquid, but this will pose no problem to perform this useful test.

1. (a) Presence of $-CO_2H$ and $-SO_3H$ groups makes the organic compounds acidic (of A_1, class). Some substituted phenols (e.g. picric acid) and salts of organic basic compounds with inorganic acids also display acidic property (i.e. evolve CO_2 with $NaHCO_3$). Phenols are considered to be weakly acidic compounds (of A_2 class) reacting with strong alkali (e.g. NaOH) to form salts (Ar–O–Na).

(b) Saturated $NaHCO_3$ contains 5% of the base in aqueous solution. The sample is taken in a clean t.t. and aqueous saturated solution of $NaHCO_3$ is added to it. If the sample is "strong" organic acid (i.e. of A_1, class) then gas bubbles (CO_2) will be evolved. If the sample is solid and insoluble in H_2O, then these solid particles stick to the CO_2 micro-bubbles and floats on the surface of water with a very few exceptions (where the CO_2H group is hindered or otherwise made inactive) all carboxy acids evolve CO_2 from $NaHCO_3$.

This CO_2 comes from the $NaHCO_3$ instead of the $-COOH$ group. If the carbon of $-CO_2H$ group is labelled (C^*) the evolved CO_2 does not contain the labelled carbon (i.e. C^*O_2 is not evolved). Likewise when $-SO_3H$ group reacts with $NaHCO_3$, only CO_2 is evolved, not SO_2 gas. Thus

$$(Ar)/R-CO_2H + NaHCO_3 \rightarrow (Ar)/R-COONa + H_2O + CO_2\uparrow$$

$$(Ar)/R-SO_3H + NaHCO_3 \rightarrow (Ar)/R-SO_3Na + H_2O + CO_2\uparrow$$

It may be noted that if Na_2CO_3 is used instead of $NaHCO_3$ then also CO_2 gas will be evolved, but now almost double the quantity of the sample should be required because the first mol. of (Ar)/R–COOH will convert the Na_2CO_3 into $NaHCO_3$ (the acid is consumed, but no $CO_2\uparrow$), then the second mol. of the acid will react with the newly formed $NaHCO_3$ to generate CO_2 which will be evolved as bubbles in the reaction mixture. Thus,

(i) $Na_2CO_3 + \underset{\text{1st mol.}}{R{-}COOH} \rightarrow R{-}CO_2Na + NaHCO_3$

(ii) $NaHCO_3 + \underset{\text{2nd mol.}}{R{-}CO_2H} \rightarrow RCO_2Na + H_2O + CO_2\uparrow$

So it is advised to take $NaHCO_3$ instead of Na_2CO_3 in this experiment.

(c) This test is known as *iodate–iodide test* which is very sensitive and can detect very minute quantity of acidic compounds. Blank test should be done with the reagents before adding the sample to the reagent.

A solution of KI (2%, 2 drops) is mixed with a solution of KIO_3 (4%, 2 drops) in a t.t. and freshly prepared starch solution (1 ml) is added to it. The mixture is shaken and if it remains colourless then this iodide–iodate–starch reagent will be suitable for use. A small amount of the sample is added to the reagent, shaken and if there is the appearance of characteristic deep violet colour of the starch–iodine complex then it would be concluded that the sample is acidic. Thus,

(i) $KI + RCO_2H \rightarrow R{-}CO_2K + HI$; (ii) $KIO_3 + R{-}CO_2H \rightarrow R{-}CO_2K + HIO_3$.

(ii) $5HI + HIO_3 \rightarrow 3I_2 + 3H_2O$; (iv) I_2 + Starch → Starch–Iodine Complex (deep-violet colour)

(d) The following equations wile explain the colour change.

(i) $Na_2CO_3 + HOH = NaOH + NaHCO_3$

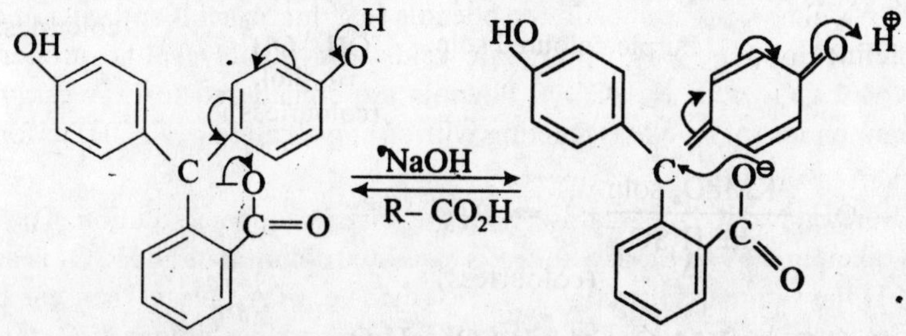

Phenolphthalein (colourless, no extended conjugation) (pink, due to long chain conjugation)

(e) The organic acid may be insoluble in H_2O but its K or Na salts should be soluble in water. Therefore, the soluble salts of organic acid on acidification with inorganic acid (which are stronger than the organic acid) will liberate the organic acid and inorganic salt will be formed. The organic acid being insoluble in H_2O will be pr cipitated out.

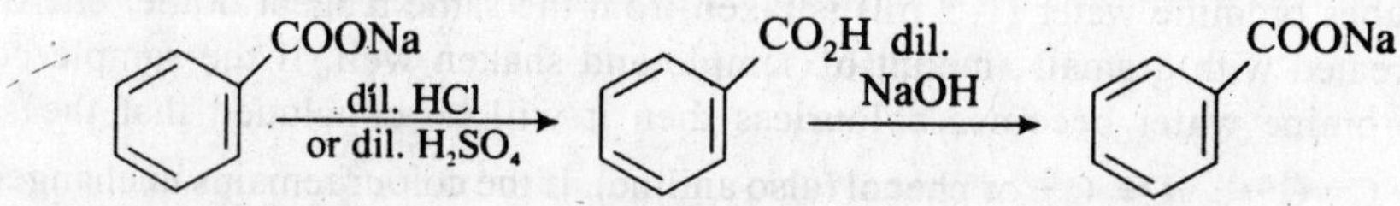

Sod. benzoate (soluble in H_2O) Benzoic acid (insoluble in H_2O) Sod. benzoate (soluble in H_2O).

Similary, the salts (soluble in H_2O) of organic bases (insoluble in H_2O) are precipitated out as the free base (insoluble) on basification with stronger inorganic bases, but again dissolve on acidification with inorganic acids.

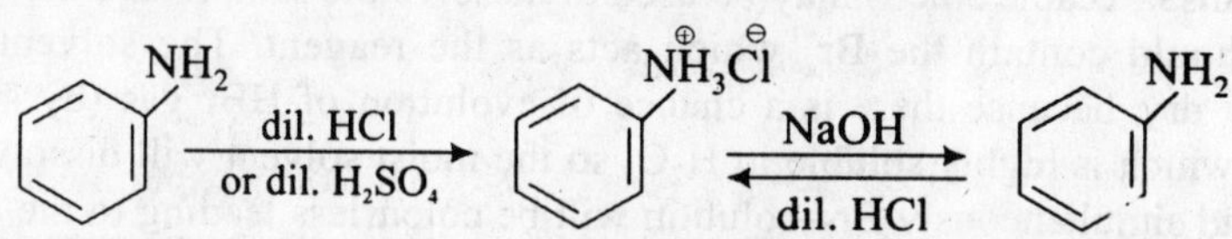

Aniline (insoluble in H_2O) Aniline hydrochloride (salt, soluble in H_2O) Aniline generated (insoluble in H_2O)

2. This is the test for alk–1–yne, the H on C_1 is acidic (50% s character) hence forms Ag salt liberating HNO_3 which turns blue litmus red. This test should be omitted when the sample is not neutral.

$R–C \equiv CH + AgNO_3 = R – C \equiv C\,Ag + HNO_3$ (It turns blue litmus red) but $R–C \equiv CR$
(1 – yne)

$+ AgNO_3 \rightarrow$ No reaction.

3(a) Vide [2.3(E)]. The following equation is illustrative.

(i) 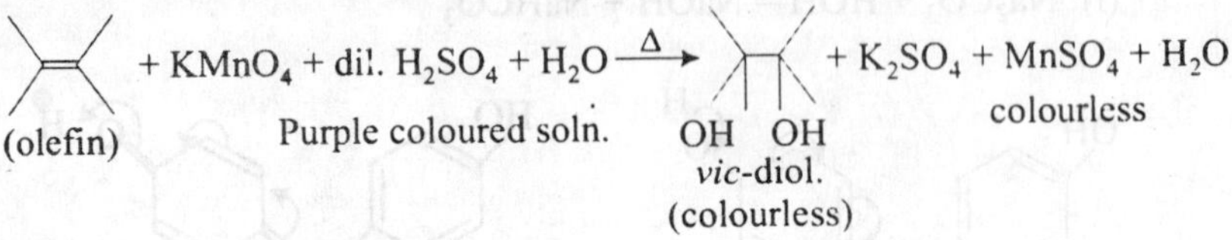

(ii) 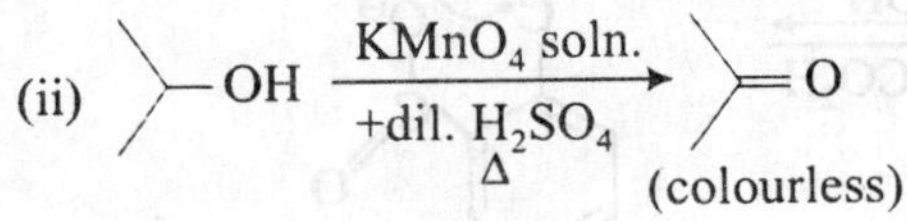

(iii) $-C\equiv C- + KMnO_4 + \text{dil. } H_2SO_4 + H_2O$
(purple colour)

$\longrightarrow -\underset{OH}{C}=\underset{OH}{C}- + K_2SO_4 + MnSO_4 + H_2O$ (colourless compounds)

(with alkynes also the $KMnO_4$ soln. becomes colourless).

(b) Elemental bromine settles at the bottom of the bottle containing bromine water. The aqueous solution of bromine (which is yellow) is used as the reagent. In two test tubes bromine water (1–2 ml) is taken from the same reagent bottle, one of them is treated with a small amount of sample and shaken well. If the sample containing bromine water becomes colourless then it will be concluded that the sample is −C=C−, −C≡C− or phenol (also aniline). If the colour remains unchanged at room temperature then both the test tubes are warmed at the same time to note if decolourisation takes place with the sample. But it must be remembered that Br_2 is a volatile element, and on strong heating both the test tubes may be colourless.

(c) If the sample is insoluble in H_2O, then any one of the solvents from CCl_4, $CHCl_3$ and CS_2 (disagreeable smell) may be used to dissolve the sample and the correspond solvent should contain the Br_2 which acts as the reagent. The solvents must be absolutely dry, because there is a chance of evolution of HBr gas (by substitution reaction) which is highly soluble in H_2O, so the moist solvent will dissolve the HBr formed and simultaneously Br_2–solution will be colourless leading to the conclusion of addition reaction resulting from the unsaturated compounds (alkenes or alkynes). Thus,

(i) 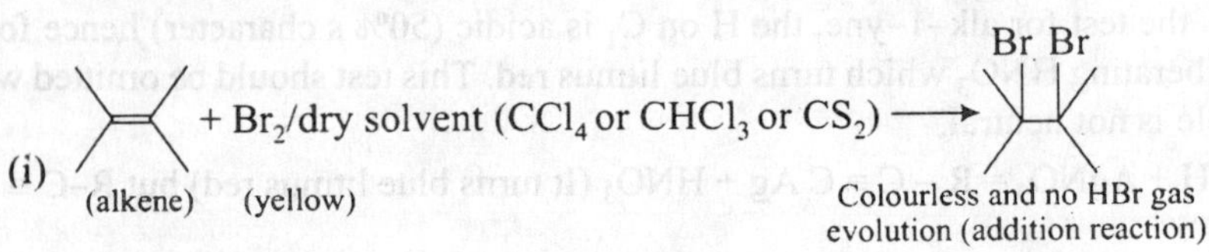

(ii) 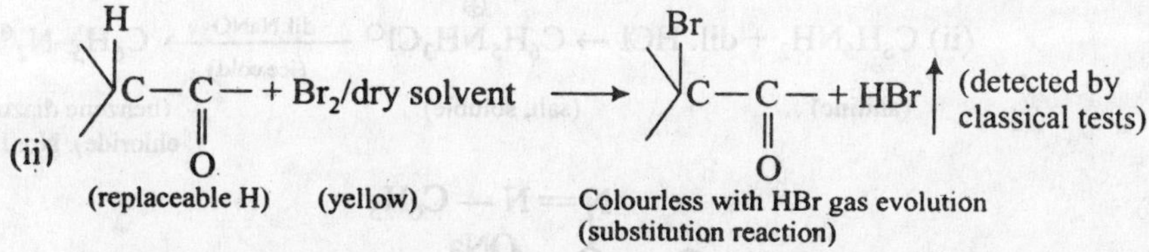

4. (a) Very dilute $FeCl_3$ solution in H_2O or EtOH (pale yellow, 0.5–1%, 3 ml) is added to a solution of the sample (2–3 drops if liquid or a pinch of the sample if solid) in the same solvent. If blue, violet or green colour appears then sample may be phenol, enol or phenolic ..ds. This resultant coloured solution is divided into two parts. To one part is added dilute HOAc, if the colour is discharged the sample is a phenol or enol. But if the colour is not discharged by HOAc then to the second part of the coloured solution is added dilute HCl when the solution becomes colourless suggesting the sample as a phenolic carboxylic acid. Thus,

(i) $6C_6H_5\text{–}OH + FeCl_3 \rightarrow H_3\,[Fe\,(O - C_6H_5)_6] + 3HCl$

(phenol, almost (pale yellow) colourless) — (coloured complex) ↓ dil. HOAc [colourless phenol regenerated].

(ii) $C_6H_4\Big]^{-OH}_{-CO_2H} + FeCl_3 \rightarrow$ coloured complex $\xrightarrow{\text{dil HCl}}$ colourless solution results

Ortho (it is salicylic acid, colourless) (yellow) — ↓ dil. HOAC. colour is not discharged (blue, green or violet)

4. (b) This is "azo-dye" test, done in the following manner:

(i) In a clean t.t. sample (2–3 drops or a pinch) is dissolved in dil. NaOH (2 ml). [Number I solution].

(ii) In another t.t. aniline (4–5 drops) is dissolved in dil. HCl (6N, 2 ml). This is cooled in ice and an ice–cold solution of $NaNO_2$ in H_2O is added to it to prepare benzene diazonium chloride (pale yellow solution). [Number II solution].

(iii) To the No. I solution is added the No. II solution (not the reverse) when a deep red solid ppt. (or a deep red solution) is obtained indicating the sample as a phenolic compound. Thus,

(i)

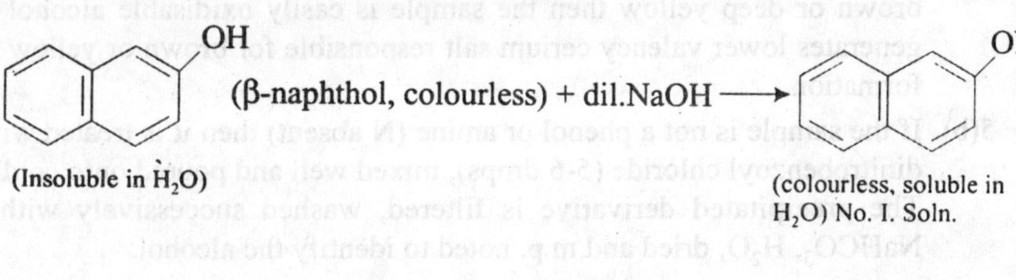

(ii) $C_6H_5NH_2$ + dil. HCl → $C_6H_5\overset{\oplus}{N}H_3Cl^{\ominus}$ $\xrightarrow[\text{(ice cold)}]{\text{dil.NaNO}_2}$ $C_6H_5{-}N_2^{\oplus}Cl^{\ominus}$

(anilinc) (salt, soluble) (benzene diazonium chloride). No. II soln.

I + II ⟶ 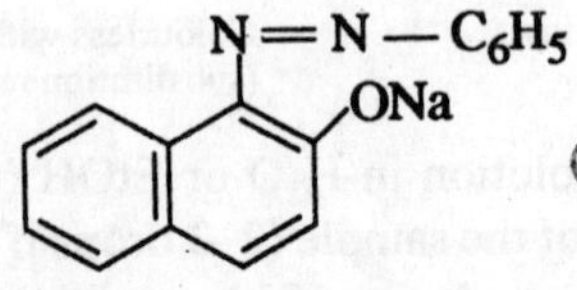 (Bright red azo-dye)

4 (c) (i) To the very dil. $FeCl_3$ solution (pale yellow) is added dil. aqueous solution of potassium ferricyanide (0.5–1%) dropwise when light brown solution of Ferric–Ferricyanide is obtained. $FeCl_3$ solution should not be added to the potassium ferricyanide to prepare the reagent.

(ii) To the solution of the sample in H_2O or EtOH is added the reagent, prepared in the first step, and shaken when deep blue ppt. is formed. Sometimes deep blue soln. may result. Phenol and aromatic amine respond to this test. It has been observed that direct addition (a very small amount, solid or liquid) of the sample to the reagent also gives the same result.

(iii) It is to be noted that picric acid does not respond although it is a phenol also aliphatic amine and enol do not respond. So phenol and enol can be distinguished by this method (only $FeCl_3$ test cannot). Many phenols (e.g. β-naphthol) do not respond satisfactorily with $FeCl_3$ solution. but with Ferric–Ferricyanide reagent immediate deep blue ppt. is obtained; This test is extremely sensitive to phenols.

(iv) It has been observed that Ferri–Ferri or Ferro–Ferro compounds are light brown but Ferri–Ferro or Ferro–Ferri are deep blue. Thus,

Phenol + Ferric–Ferricyanide $\xrightarrow[\text{of Phenol}]{[O]}$ Ferro–Ferricyanide or Ferric Ferrocyanide deep blue (Prusian blue)

5(a) Ceric ammonium nitrate (i.e. CAN) is dissolved in H_2O (a light yellow soln. is made from ca 100 mg of CAN) and divided into two parts (1–2 ml each part). To one part is added the sample and both the parts are shaken comparing the change of colour. If the sample can turn the CAN solution brown or deep yellow then the sample is easily oxidisable alcohol which generates lower valency cerium salt responsible for brown or yellow colour formation.

5(b) If the sample is not a phenol or amine (N absent) then it is treated with 3,5-dinitrobenzoyl chloride (5-6 drops), mixed well and poured onto cold water. The precipitated derivative is filtered, washed successively with H_2O, $NaHCO_3$, H_2O, dried and m.p. noted to identify the alcohol.

The reagent is prepared just before use by taking in a mortar 3,5-

dinitrobenzoic acid (ca 1 gm) and treating it with PCl_5 when the reaction starts vigorously with evolution of lots of fumes. It is wise to perform this step near an exhaust fan or in a fume chamber so that the HCl gas, thus emitted, cannot pollute the laboratory atmosphere. This reaction mixture is ground well by a pestle and the 3,5-dinitrobenzoyl chloride, thus prepared, is ready for use.

(i) 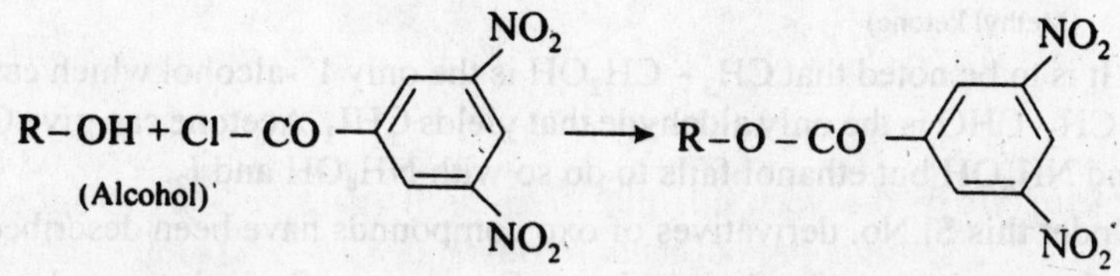

(Alkyl 3,5-dinitrobenzoate)

(ii) 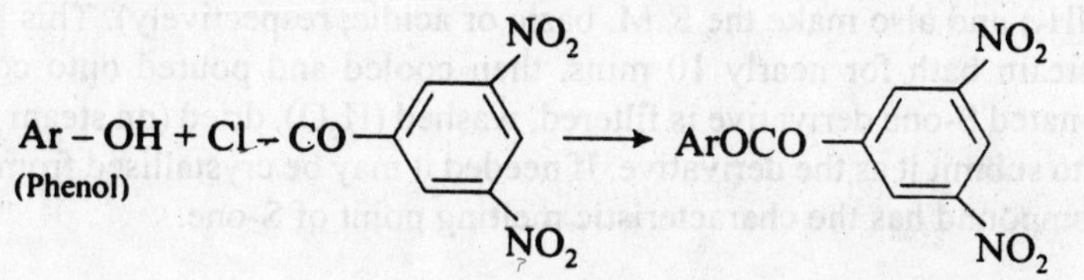

(Aryl 3,5-dinitrobenzoate)

(iii) 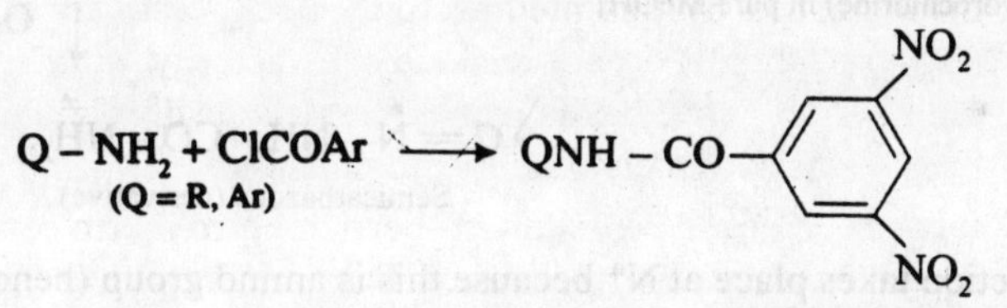

(3,5-dinitrobenzoyl derivative of the amine)

5. (c) In an Erlenmeyer flask (i.e. conical flask) the sample (ca 10 drops) is taken and H_2O (20 ml) added to it followed by I_2 powder (ca 5 gm) and warmed to nearly 70°C. Dil. NaOH soln. (1N) is added dropwise to this mixture with stirring until all iodine disappear with the simultaneous deposition of heavy CHI_3 with characteristic smell and yellow colour. CHI_3 paralyses the olfactory nerve, so repeatedly taking smell of the yellow product to confirm its identity is useless. The CHI_3, thus obtained, is filtered, washed with plenty of H_2O (to remove sticky NaOH), dried, m.p. noted (m.p. 119°) and if

directed may be submitted as the derivative of the supplied sample.

(i) $R(\text{or } Ar){-}CH(OH){-}CH_3 + 4I_2 + 6NaOH = R/Ar{-}CO_2Na + 5NaI + 5H_2O + CHI_3$

(Methyl carbinol) (iodoform)

$\downarrow I_2 + 2\,NaOH$

(ii) $R(\text{or } Ar){-}CO{-}CH_3 + 3I_2 + 4\,NaOH = R/Ar - CO_2Na + 3NaI + 3H_2O + CHI_3$.

(Methyl ketone)

(iii) It is to be noted that $CH_3 - CH_2OH$ is the only 1°-alcohol which can furnish CHI_3 and $CH_3{-}CHO$ is the only aldehyde that yields CHI_3. Acetone can give CHI_3 even with I_2 and NH_4OH but ethanol fails to do so with NH_4OH and I_2.

6. Under this Sl.No. derivatives of oxo compounds have been described.

6(a) *Semicarbazone (= S-one):* In an Erlenmeyer flask the sample in acetone-free methanol is taken, semicarbazide hydrochloride is added to it followed by either pyridine (C_5H_5N) or NaOAc (they remove HCl and set free semicarbazide ($H_2N{\cdot}NH{\cdot}CO{-}NH_2$) and also make the R.M. basic or acidic, respectively). This R.M. is heated on a steam bath for nearly 10 mins, then cooled and poured onto cold water. The precipitated S-one derivative is filtered, washed (H_2O), dried (on steam bath) and m.p. noted to submit it as the derivative. If needed it may be crystallised from ethanol. Each oxo compound has the characteristic melting point of S-one.

$$HCl{\cdot}H_2\overset{*}{N} - NH - CO - \overset{\neq}{N}H_2 + C_5H_5N \text{ or } NaOH \longrightarrow H_2\overset{*}{N} - NH - CO - \overset{\neq}{N}H_2$$

[(Sem. hydrochloride) in pure MeOH]

$$\downarrow O\underset{\Delta}{=}C\langle \text{ (aldehyde/ketone)}$$

$$\rangle C = \overset{*}{N} - NH - CO - \overset{\neq}{N}H_2,$$

Semicarbazone (derivative)

The reaction takes place at N* because this is amino group (hence basic), and not at N≠ because this is amide nitrogen ($-CON^{\neq}H_2$) hence neutral, so it cannot react with the carbonyl group to yield the S-one.

6(b) As in (a) if semicarbazide hydrochloride is replaced by hydroxylamine hydrochloride (these active reagents are unstable so they are made stable as hydrochlorides, i.e. salts of inorganic acids), then oxime derivatives of oxo-compounds will be formed.

$$HO - NH_2{\cdot}HCl + O = C\langle \xrightarrow[\text{NaOAC/MeOH, }\Delta]{\text{MeOH,}C_5H_5N,\ \Delta \text{ or}} HO - N = C\langle$$

hydroxylamine hydrochloride — Oxime derivatives, have characteristic melting points

6(c) Similarly, when 2,4-dinitrophenylhydrazine hydrochloride (or sulphate) reacts with the oxo-compound then 2,4-dinitrophenylhydrazone (i.e. 2,4-DNP) derivative is obtained. The derivative can be prepared and purified very easily hence the 2,4-DNP derivative is the choice in most of the cases.

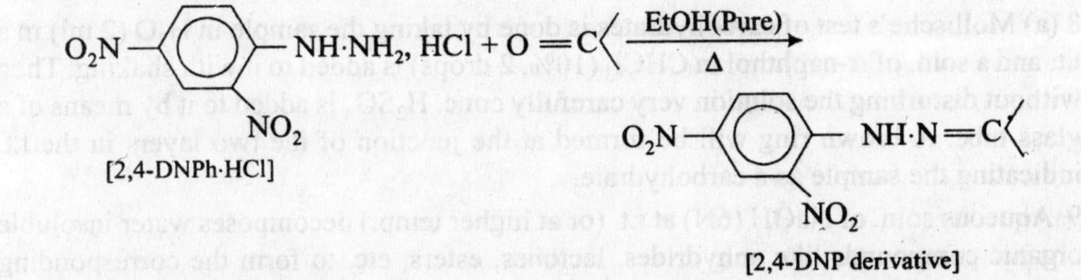

7(a) The red colour of fuchsine (i.e. magenta dye) in aqueous solution is discharged by passing SO_2 gas through it. This colourless solution is Schiff's reagent which is treated with a small amount of the sample and warmed. If the red colour reappears then the sample is an aldehyde. If the sample is a ketone then the red colour of fuchsine will not reappear.

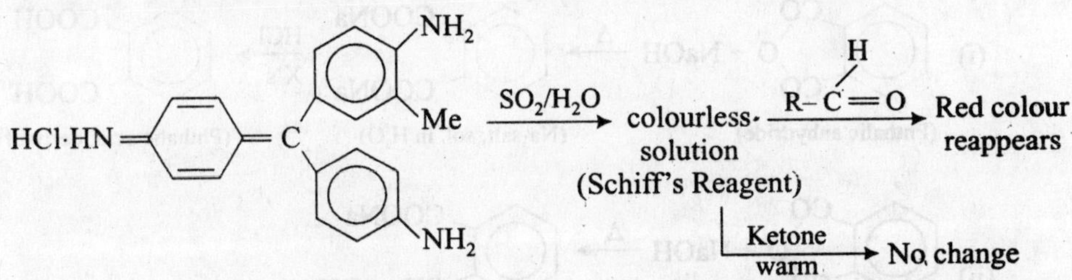

7(b) Fehling's solution A (or I) is aqueous soln. of $CuSO_4 \cdot 5H_2O$ with a few drops of H_2SO_4, which prevents the decomposition of $CuSO_4$, thus making the "Soln. A" stable. Fehling's solution B (or II) is aqueous soln. of sodium potassium tartrate made strongly alkaline with NaOH solution. Equal volumes of solutions. A and B are mixed together to form deep-blue copper complex which is used to identify aldehydes which generates red Cu_2O as product. Ketones do not respond to this test if boiled for a short time. Benedict's soln. also responds similarly.

(i) $R\text{–}CHO + 2Cu^{2\oplus} + 5\,OH^{\ominus} = R\text{–}CO_2^{\ominus} + Cu_2O\downarrow + 3\,H_2O$

(ii) $R\text{–}CHO + 2\,CuSO_4 + 5NaOH = R\text{–}CO_2Na + 2Na_2SO_4 + Cu_2O\downarrow + 3\,H_2O$.

7(c). To the solution of $AgNO_3$ in H_2O is added a drop of NaOH when black ppt. is formed which is treated with dil. NH_4OH (dropwise) with shaking. The addition of NH_4OH should be stopped as soon as the black ppt. dissolves. This is Tollens reagent which furnishes black Ag powder as ppt. on heating on a steam bath with an aldehyde. During heating the R.M. on a steam bath if the container (i.e. the t.t.) kept still then silver mirror (shining layer) is formed. Ketones do not react with Tollens reagent. This reagent should never be heated on a direct flame. The stock soln. of the reagent is never made. Thus,

(i) $R\text{–}CH = O + 2Ag\,(NH_3)_2^{\oplus} + 3\,HO^{\ominus} = R\text{–}COO^{\ominus} + 2Ag\downarrow + 4NH_3 + 2H_2O$.

(ii) $H\text{–}CH = O + 4\,Ag\,(NH_3)_2^{\oplus} + 4\,HO^{\ominus} = CO_2 + 4\,Ag\downarrow + 8\,NH_3 + 3\,H_2O$.

8 (a) Mollische's test of carbohydrates is done by taking the sample in H_2O (2 ml) in a t.t. and a soln. of α-naphthol in $CHCl_3$ (10%, 2 drops) is added to it with shaking. Then without disturbing the solution very carefully conc. H_2SO_4 is added to it by means of a glass tube. A brown ring will be formed at the junction of the two layers in the t.t. indicating the sample as a carbohydrate.

9. Aqueous soln. of NaOH (6N) at r.t. (or at higher temp.) decomposes water insoluble organic compounds like anhydrides, lactones, esters, etc. to form the corresponding Na–salts which are soluble in water. Thus generally at the end of the reaction a transparent clear solution is obtained. But if the sample is an ester of an alcohol which is water insoluble (e.g. benzyl benzoate, iv), then the product will not be in a transparent solution. In most of the cases the clear soln. on acidification with inorganic acids will generate the acidic part (from the Na–salt) which may be collected as derivative.

(i) 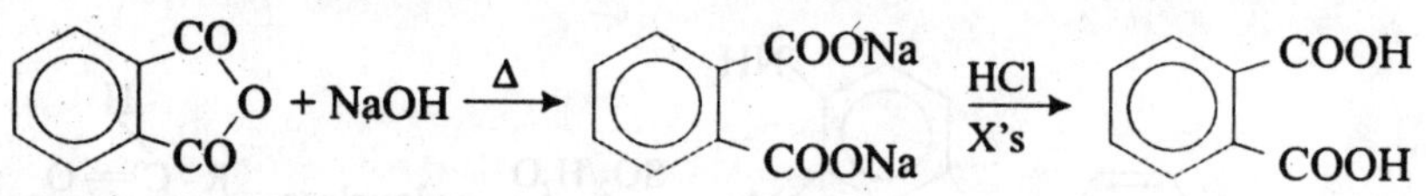

(Phthalic anhydride) (Na-salt, sol. in H_2O) (Phthalic acid, derive.)

(ii) 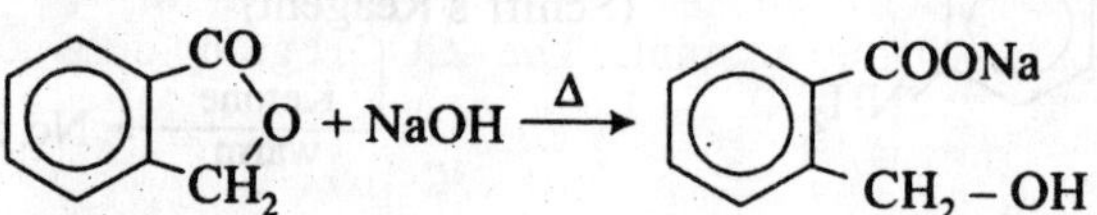

(Lactone)

(iii) $$\underset{\text{(Ester)}}{C_6H_5\text{–}COOC_2H_5} + NaOH \xrightarrow{\Delta} \underset{\text{Soluble}}{\left[C_6H_5\text{–}COONa + EtOH\right]} \xrightarrow[\text{HCl}]{\text{conc.}} \underset{\text{(Benzoic acid, deriv.)}}{C_6H_5COOH}$$

(iv) $$\underset{\text{(Benzyl benzoate)}}{C_6H_5\text{–}COO\text{–}CH_2\text{–}C_6H_5} + NaOH \xrightarrow{\Delta} \left[\underset{\text{Soluble}}{C_6H_5 - COONa} + \underset{\text{Insoluble}}{HOCH_2 - C_6H_5}\right]$$

(The R.M. is not clear)

$$\xrightarrow[\text{HCl}]{\text{conc.}} \text{Benzoic acid, deriv.}$$

10. Hydroxamic acid test for ester is done by taking in a t.t. the sample (0.5 ml) and very small amount of hydroxylamine hydrochloride (excess to be carefully avoided) followed by dil. NaOH soln. to make the mixture alkaline (the reaction fails in acidic medium). If required to make the R.M. clear a few drops of EtOH may be used. The R.M. is boiled for 5 mins. when it becomes light brown solution (if not the boiling should be continued, e.g. with diethyl phthalate) which is cooled, acidified with dil. HCl (1N) and treated with very dilute $FeCl_3$ soln. (pale yellow). The appearance of purple, violet or brown colour indicates the sample as ester. On very rare occasions

acid chloride and anhydride respond to this test. In is noteworthy that acetanilide responds to this test, moreover phenols do not interfere with this test since this is carried out in dil. HCl medium which destroys the purple colour caused by reaction of phenol with $FeCl_3$. Thus,

(i) $\underset{\text{(Phenol)}}{C_6H_5OH} + \text{dil. } FeCl_3 \rightarrow \text{Purple colour} \xrightarrow[\text{HCl}]{\text{dil.}} \underset{\text{(colourless)}}{\text{Phenol}}$

(ii) $\underset{\text{(Ester)}}{C_6H_5\text{–}CO_2R} + HONH_2{\cdot}HCl + NaOH \xrightarrow{\Delta} \underset{\text{(alkaline)}}{\text{Soln.}} \xrightarrow[\text{2) } FeCl_3]{\text{1) dil. HCl}} \underset{\text{(stable in dil. HCl).}}{\text{Violet colour soln.}}$

(iii) $C_6H_4\,]^{-OH}_{-CO_2CH_3} + HONH_2{\cdot}HCl + NaOH \xrightarrow{\Delta} \underset{\text{(alkaline)}}{\text{Soln.}} \xrightarrow[\text{2) } FeCl_3]{\text{1) dil. HCl}} \underset{\text{(Stable in HCl)}}{\text{Purple colour soln.}}$

ortho (Methyl salicylate)

$\xrightarrow{\text{dil. } FeCl_3}$ Purple colour (due to phenolic OH) $\xrightarrow{\text{dil. HCl}}$ Colour is discharged (Phenol-$FeCl_3$ colour is unstable in dil. HCl)

The example (iii) shows that if both phenolic –OH and –CO_2R groups are present, then both of them can be identified in a sample. The –COOH group does not interfere.

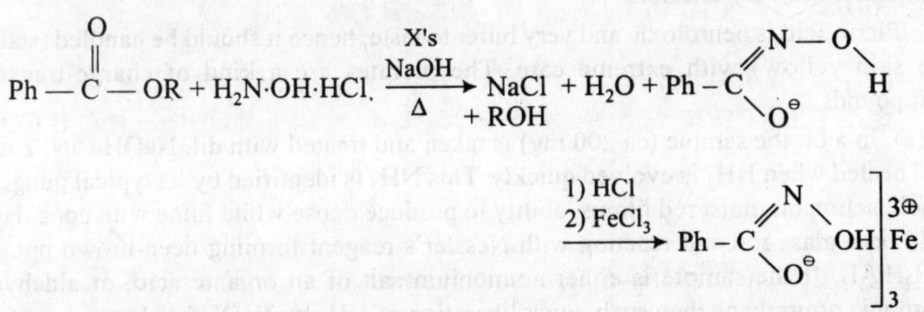

(Violet/Purple)

11. A mixture of constant boiling HI (b.p. 127°C, 57% HI, freshly distilled) and aromatic/aliphatic ether (Ar–OR or R′–OR, R < Ar/R′) is heated under reflux. The ether decomposes into Ar–OH or R′–OH and RI is formed (since R<Ar/R′; the smaller alkyl group will form the more volatile alkyl iodide). This RI will finally react with $AgNO_3$ but before that the accompanying HI should be trapped by passing the mixture of RI and HI through the suspension of red phosphorus in H_2O which retains HI. The RI, thus washed free from HI is passed through $AgNO_3$, and the yellow ppt. of AgI indicates the sample as ether.

If the sample is an aromatic ether, then the phenol (Ar–OH) formed can be converted into suitable derivatives and its identity can be established. This phenol will also respond to ferric–ferrocyanide test mentioned before:

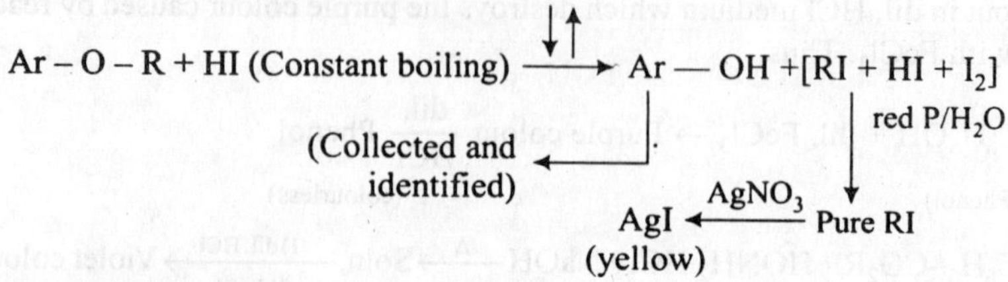

12. When no functional group has been detected then this test should be done. Generally, aromatic hydrocarbons furnish insoluble picrates (as derivatives) but other compounds like phenols, amines, etc. also respond to this test.

Picric acid (2,4,6-trinitrophenol) is an explosive compound when dry, hence this is always kept under water. Just before use it is taken out of water, dried over blotting paper or filter paper (not recommended), weighed mole for mole with the substrate and dissolved in either pure EtOH or benzene using the same solvent to dissolved the substrate. Generally, 0.001 mole of the picric acid is reacted with 0.001 mole (exact proportions) of the substrate. The R.M. is heated on a steam bath for a few minutes, cooled and the precipitated picrate is filtered, dried and m.p. noted of the derivative. If the picrate is soluble in the solvent, then on a porcelain basin the R.M. is taken and the solvent is evaporated off to collect the picrate as soon as possible because the picrates are sometimes very unstable.

Picric acid is neurotoxic and very bitter to taste, hence it should be handled (stains the skin yellow) with extreme care. The picrates are a kind of charge-transfer compounds.

13(a). In a t.t. the sample (ca 200 mg) is taken and treated with dil. NaOH (6N, 2 ml) and heated when NH_3 is evolved quickly. This NH_3 is identified by its typical pungent smell, action on moist red litmus, ability to produce dense white fume with conc. HCl held on a glass rod and reaction with Nessler's reagent forming deep-brown ppt. of $NH_2Hg_2I_3$. If the sample is either ammonium salt of an organic acid, or aldehyde ammonia or urethane then such quick liberation of NH_3 by NaOH has been observed.

(i) $R\text{–}COONH_4 + NaOH \xrightarrow{\Delta} RCO_2Na + H_2O + NH_3\uparrow$

(ii) $R\text{–}CH(OH)NH_2 + NaOH \xrightarrow{\Delta} R\text{–}CHO + H_2O + NH_3\uparrow$

13(b). In an Erlenmeyer flask (100 ml) the sample (200 mg) is taken, treated with dil. NaOH (6N, 10 ml) and heated under reflux when NH_3 is evolved quickly [vide 13 (a)] when the sample is an acid amide. But if NH_3 is evolved comparatively slowly then the sample should be a nitrile.

(i) $R\text{–}CO\text{–}NH_2 + HO^{\ominus} \xrightarrow{\uparrow\downarrow} R\text{–}CO_2^{\ominus} + NH_3\uparrow$

(ii) $R–CN + HO^{\ominus} + H_2O \xrightarrow{\uparrow\downarrow} R–CO_2^{\ominus} + NH_3\uparrow$

13(c). The sample (small quantity) was taken in a t.t. and treated with dil. NaOH. The R.M. is warmed a little and treated with excess of a mixture of dil. $AgNO_3$ and dil. HNO_3 until the R.M. is acidic. A white ppt. of AgCN (insoluble in HNO_3, but soluble in NH_4OH, cf. AgCl.CN; = ψ halogen) indicates the sample as cyanohydrine.

(i) $R–CH\,(OH)\,CN + NaOH \rightarrow NaCN \xrightarrow[HNO_3]{AgNO_3} AgCN\downarrow$ white ppt.

14(a). The amines (1°, 2° and 3°) are basic, hence soluble in dil. HCl forming salts (hydrochlorides). These salts are slightly acidic due to the influence of HCl being stronger than the corresponding organic bases.

(i) $\underset{\text{salt. (soluble in } H_2O)}{R–NH_2 + HCl \rightarrow R–N^{\oplus}H_3Cl^{\ominus}} \overset{HOH}{\rightleftharpoons} \underset{\text{weak base}}{R–NH_2} + \underset{\text{strong acid.}}{HCl} \text{ (aq. soln. acidic)}.$

In some cases it has been observed that the resultant hydrochlorides are not soluble in H_2O (e.g. β-naphthylamine) but on dilution with large volume of H_2O these salts dissolve and responds to all the tests of amines (β-naphthylamine is a carcinogenic compound).

14(b) In a t.t. the sample (100 mg) and dil. HCl (1N, 4 ml) are taken and cooled in ice and treated with ice-cold dil. solution of $NaNO_2$ and slowly allowed to attain the room temperature. If the sample is 1° aliphatic amine ($R–NH_2$), then it will be diazotised in cold but will decompose at room temperature to evolve N_2 gas. On the contrary 2° and 3° amines will furnish oily or solid ppt.

14(c). In a t.t. the sample (100 mg) is dissolved in dil. HCl (1N, 4 ml), cooled in ice, treated with dil. ice-cold $NaNO_2$ solution and added to the β-naphthol (100–200 mg) dissolved in dil. NaOH (6N, 6 ml). A brilliant red azo-dye formation will indicate 1° aromatic amine (vide 4(b)). It should be noted that some anilide (e.g. acetanilide) undergoes ready hydrolysis with dil. HCl even at room temperature, so caution must be exercised before final conclusion.

14(d). If the sample does not respond to 14(c), then the sample (100 mg) is treated with conc. HCl (5 ml) and the mixture is heated under reflux until a clear solution is obtained. This R.M. is diluted with water, cooled in ice, treated with ice-cold $NaNO_2$ solution, then poured onto alkaline solution of β-naphthol (in t.t.) and the formation of red azodye indicates the sample as anilide (Ar NH–COR), Bezanilide ($C_6H_5NH–CO–C_6H_5$) is very resistant to hydrolysis with conc. HCl, so more drastic condition is taken to hydrolyse the compound by refluxing with 70% H_2SO_4 (mixture of equal volumes of H_2SO_4 and H_2O).

(i) $\underset{(R \neq C_6H_5)}{Ar\,NH–COR} \xrightarrow[\uparrow\downarrow\,5\text{ mins}]{\text{Conc. HCl, } + H_2O} ArNH_2{\cdot}HCl \xrightarrow[NaNO_2]{\text{ice cold}} ArN^{\oplus}_2Cl^{\ominus}$

$\xrightarrow[\beta\text{-naphthol}]{\text{alk.}}$ Azo-dye.

(ii) $Ar–NH–COC_6H_5 \xrightarrow[70\%\ H_2SO_4\,\uparrow\downarrow]{} \underset{(+\,C_6H_5COOH)}{(ArNH_2){\cdot}H_2SO_4} \xrightarrow{-do-}$

$\downarrow$

$ArN^{\oplus}_2 \xrightarrow{-Do-}$ Azodye

14(e). If Ar–NH_2 and Ar–NH–COR are both absent [14(c) and 14(d)] then the sample may be Ar–NO_2 which on reduction will generate Ar–NH_2. This derivative should be identified by test 14(c). Hence the sample (500 mg) is taken in an Erlenmeyer flask, a few pieces of granulated tin are added to it followed by conc. HCl (10 ml) and the mixture is heated under reflux until the insoluble oily nitro compound dissolves completely as $ArNH_2{\cdot}HCl$. Now the R.M. is diluted with H_2O (50 ml), cooled in ice, diazotised with ice-cold $NaNO_2$ soln. and coupled with alkaline β-naphthol. The formation of red azo-dye indicates aromatic nitro compounds.

$$\underset{\text{(Aromatic nitrocompd.)}}{ArNO_2} + Sn + \text{Conc. HCl} \xrightarrow{\uparrow\downarrow} ArN^{\oplus}H_3Cl^{\ominus} \xrightarrow[14(c)]{\text{ice cold}} \text{Red azo-dye.}$$

14(f). This is a special test for Ar–NO_2 but not very illustrative.

The sample (50 mg) is taken in a t.t. and mixed with dil. NaOH (6N, 1 ml). On heating if the mixture turns brown or deep yellow then it would be concluded that the sample is an aromatic nitro compound.

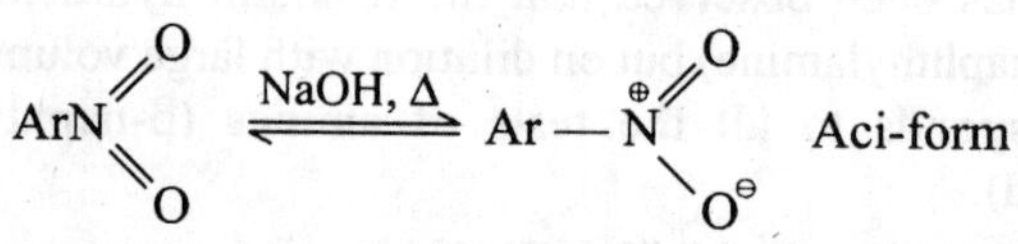

(Pale yellow) (Yellow colour becomes more intense)

14(g). If the sample contains both –NH_2 group and –NH–COR on an aromatic nucleus, then the anilide group can be identified only after removal of the –NH_2 group. So the –NH_2 is identified by 14(c), then the –NHCOR could be identified in the following way.

The sample (200 mg) is taken in a t.t., dissolved in dil. HCl, diazotised and divided into two parts. One part was tested for azo-dye, to establish the presence of –NH_2 on an aromatic nucleus, and the other diazotised part was treated with ethanol, boiled (to decompose diazonium group), conc. HCl added and boiled to hydrolyse –NHCOR into –NH_2 group. This new born –NH_2 group was again diazotised in the usual way and coupled with β-naphthol (in the alkaline medium) to form azo-dye. The reactions are shown below,

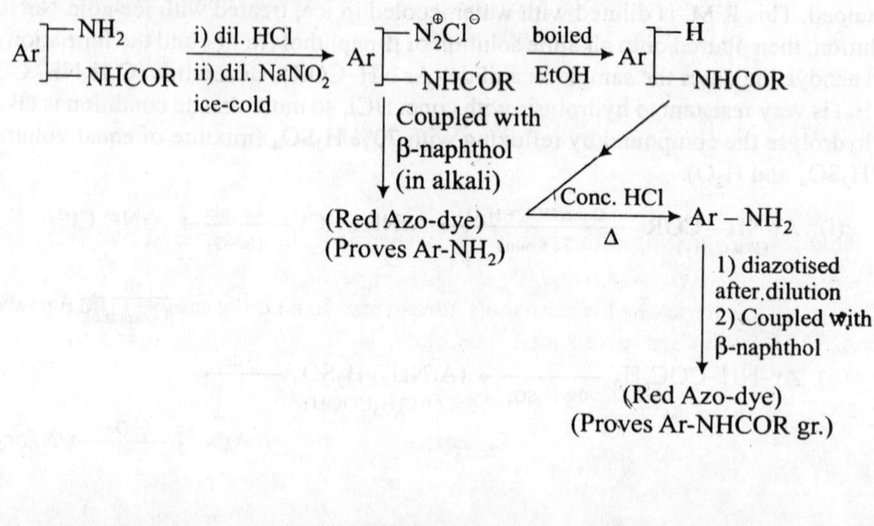

14(h). On a white sheet of paper a watch glass is placed and the sample (very small amount) is taken on it. Two drops of conc. H_2SO_4 is added on the sample and the R.M. is mixed well by a glass rod. On addition of a pinch of $K_2Cr_2O_7$ to this mixture if a rose–red colour develops then this will indicate the sample as anilide. This test is of minor importance and may be done as a complementary test of 14(d).

14(i). If the sample contains aromatic 1°–NH_2 group but NH–COR is absent then the sample may be [I] as shown below. The compound is identified in the following way.

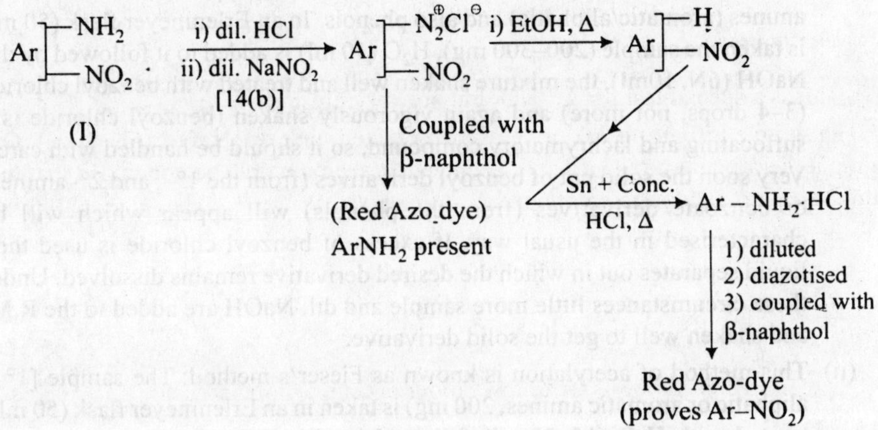

14(j). The aromatic amino-nitro compound (I) is taken in a t.t. and mixed with Zn powder followed by NH_4Cl and EtOH. The mixture is very carefully boiled (Fire, EtOH present) on a low flame (better on a steam bath) for 5 mins, cooled, filtered and the filtrate treated with Tollens reagent. Formation of silver mirror or precipitate of metallic silver powder (black) indicates the presence of $ArNO_2$ group.

14(k). If the sample contains N (element detection test) then this test should be always carried out irrespective of the presence of other functional group(s).

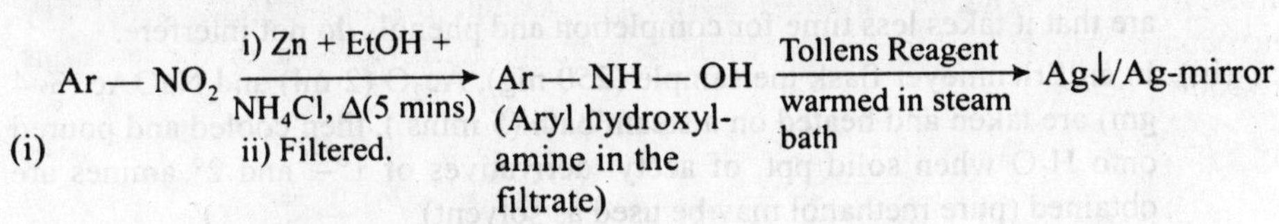

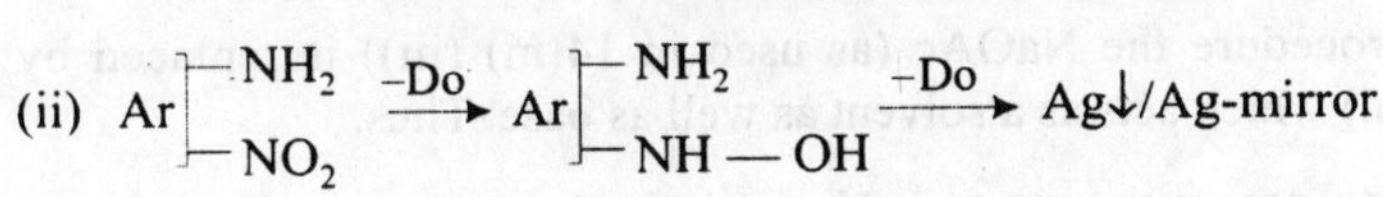

(I)

This test in known as the Mulliken–Barker Test, not interfered by –NH_2 and –NHCOR on aromatic ring.

14(l). This is known as the Liebermann's nitroso test. In a t.t. the sample (100 mg) and phenol (2–3 drops) are mixed well then conc. H_2SO_4 (0.5 ml) is added to it with shaking. The R.M. is poured onto cold H_2O and made alkaline with dilute NaOH. Formation of deep blue solution indicates the sample as nitroso-amine ($R_2N{\cdot}NO$ or Ar–N(NO)R). If the sample is 2°–amine (R_2NH), then it is treated with ice-cold HNO_2 ($NaNO_2$ + dil. HCl) to make $R_2N{\cdot}NO$ which responds to Liebermann test in the next step.

14(m)

(i) This test is known as Schotten–Baumann reaction, applicable to 1°–and 2°–amines (aromatic/aliphatic) and also phenols. In an Erlenmeyer flask (50 ml) is taken the sample (200–300 mg), H_2O (10 ml) is added to it followed by dil. NaOH (6N, 10ml), the mixture shaken well and treated with benzoyl chloride (3–4 drops, not more) and again vigorously shaken (benzoyl chloride is a suffocating and lachrymatory compound, so it should be handled with care). Very soon the solid ppt of benzoyl derivatives (from the 1°–, and 2°–amines) or benzoate derivatives (from the phenols) will appear which will be characterised in the usual way. If excess of benzoyl chloride is used then liquid separates out in which the desired derivative remains dissolved. Under these circumstances little more sample and dil. NaOH are added to the R.M. and shaken well to get the solid derivative.

(ii) This method of acetylation is known as Fieser's method: The sample [1° – aliphatic or aromatic amines, 200 mg) is taken in an Erlenmeyer flask (50 ml), treated with H_2O (15–20 ml) followed by dil. HCl (6N) dropwise to just dissolve the amine. In a t.t. fresh Ac_2O (6–7 drops) is taken and on a piece of paper NaOAc (ca 2 g) is kept ready. To the dissolved amine hydrochloride the Ac_2O is added and immediately the NaOAc is added to the R.M. which is shaken vigorously in the stoppered Erlenmeyer flask. Almost immediately the acetyl derivative appears as colourless solid which is collected in the usual way. A few aromatic amines (e.g. *o*–toluidine) do not furnish solid derivatives if excess of Ac_2O and NaOAc are used because the acetyl derivative are soluble in excess of acetic acid. The advantages of this method are that it takes less time for completion and phenols do not interfere.

(iii) In an Erlenmeyer flask the sample (250 mg), Ac_2O (2 ml) and NaOAc (3–4 gm) are taken and heated on a steam bath (5 mins.), then cooled and poured onto H_2O when solid ppt. of acetyl derivatives of 1°–, and 2° amines are obtained (pure methanol may be used as solvent).

(iv) In another procedure the NaOAc (as used in 14(m) (iii)) is replaced by pyridine (2 ml) which acts as a solvent as well as base. Thus.

(i) $$R\text{–}NH_2 + NaOH + C_6H_5\text{–}COCl \xrightarrow[H_2O]{r.t.} R\text{–}NH\text{–}COC_6H_5 + NaCl + H_2O$$

(benzoyl deriv.) Schotten–Baumann method.

$$R_2NH + NaOH + C_6H_5COCl \xrightarrow{do} R_2N\text{–}CO\text{–}C_6H_5 + NaCl + H_2O$$

$$\text{(ii)}\ RNH_2 + HCl \rightarrow R\text{–}\overset{\oplus}{N}H_3Cl^{\ominus} \xrightarrow[\substack{+NaOAc \\ (H_2O)}]{Ac_2O} R\text{–}NH\text{+}CO\text{–}CH_3 + NaCl + HOAc$$

(acetyl deriv) Fieser's method.

$$\text{(iii)}\ RNH_2\ (\text{or } R_2NH) + Ac_2O + NaOAc \xrightarrow[\text{bath, 5 min}]{\text{Steam}} R\text{–}NH\text{–}COCH_3$$

or $R_2N\text{–}COCH_3$.

$$\text{(iv)}\ R\text{–}NH_2\ (\text{or } R_2'NH) + Ac_2O + \underset{(\text{pyridine})}{C_5H_5N} \xrightarrow{\text{do}} R\text{–}NH\text{–}COCH_3$$

or $R_2N\text{–}COCH_3$.

14(n)

(i) This is a test for 3°-amine. The sample is taken in an Erlenmeyer flask (25 ml) and dissolve in ether. A few drops of CH_3I is added to this solution and shaken well when solid metho-iodide derivative is precipitated. It should be done very carefully and away from fire because highly combustible ether is employed as a solvent, the vapour is also heavy, so comes downward from the steam bath. Sometimes the metho-iodide is soluble in H_2O, so anhydrous ether should be used.

$$R_3N + CHI_3 + \text{Ether} \xrightarrow{\text{r.t. shaking}} \underset{(\text{quaternary salt})}{R_3\overset{\oplus}{N}CH_3I}\ (\text{metho-iodide}).$$

(ii) Sometimes 3°–amines reacts with aqueous solution of $K_4Fe(CN)_6$ to form solid precipitate (occasionally used to detect alkaloids).

(iii) This is a very sensitive test for 3°–aromatic amines. The characteristic deep blue or brown colour is obtained due to the formation of thionene dye.

15. 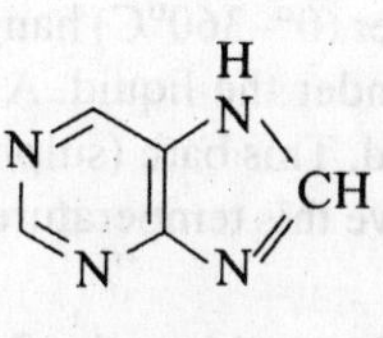
(Purine)

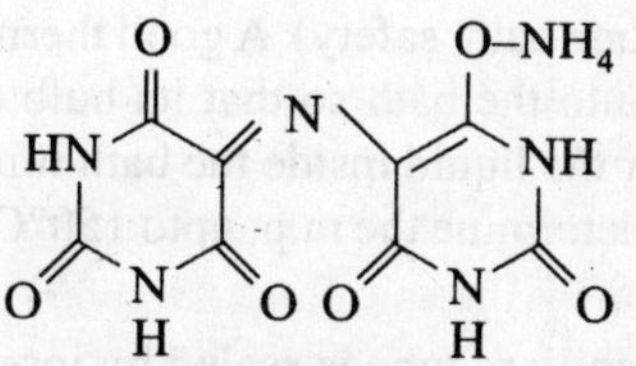
(Murexide)

16 (i) Mayer's reagent is prepared as (HgI_2 + KI = $KHgI_3$). In aqueous solution it reacts with the alkaloids to give precipitate.

(ii) Wagner's reagent is KI_3 prepared from KI and I_2. The aqueous solution is used to detect alkaloids through precipitate formation.

(iii) Dragendorff's reagent is $KBiI_4$ from KI + BiI_3. The alkaloids are precipitated in aqueous solution.

It should be noted that in addition to the alkaloids many other compounds may respond to these reagents. So when the precipitate is obtained by using these reagents the conclusion should be "Alkaloids may be present."

17. Tests (a), (b) and (c) are self-explanatory, only to be remembered that hydrohalide salts of organic bases are soluble in H_2O making the solution acidic (litmus test) and gives ppt. with a mixture of dil. $AgNO_3$ and HNO_3. These silver halides are identified in the usual way to determine the nature of the hydrohalic acid making the salt with the organic base (cf.d).

18. (a) When the sample is R(Ar)–CO–Hal then it undergoes hydrolysis by hot H_2O to generate H–Hal (i.e. HX) and organic acid. This HX is characterised by usual method (dil. $AgNO_3$ + HNO_3 mixture).

(b) Alkyl halide and aryl halide behave in the following way:

(i) $C_6H_5–CH_2–CH_2–Cl \xrightarrow[\Delta]{AgNO_3/HOH} AgCl\downarrow + Ph–CH_2–CH_2–OH$
(aliphatic halo-compd.)

(ii) 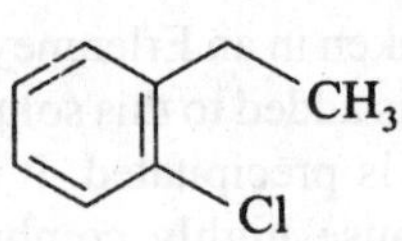 $\xrightarrow{Do}$ No reaction (no AgCl↓).
(aromatic ring holds halogen)

(c) (i) $R–CH_2\,I$ + absolute EtOH + KOH $\xrightarrow{\Delta}$ KI (no ppt.; soluble in dry EtOH).

(ii) $R–CH_2X \xrightarrow{-do-} KX\downarrow$ (ppt., insoluble in absolute EtOH).
(X = Cl/Br) (X = Cl or Br)

(iii) ArI $\xrightarrow{-do-}$ No ppt. (because the I is on aromatic ring).

(VII) *Setting up a melting point bath:* On a tripod stand iron-wire gauze is placed over which an Erlenmeyer flask (Pyrex glass in preferred, otherwise Sigcol glass, 100 ml) is clamped and almost half-filled with conc. H_2SO_4 (sulpuric acid bath, danger) or silicone oil (silicone bath, safety). A good thermometer (0°–360°C) hanging from a string is immersed into the bath so that its bulb dips under the liquid. A thin glass rod may be used to stir the liquid inside the bath if required. This bath (sulphuric or silicone) may be used to determine the m.p. upto 150°C. Above this temperature use of a metal bath is advisable.

The wider end of the capillary tube is sealed by inserting this end into the flame. The sample is powdered (on a clean watch glass) by the help of a spatula and the open end of the capillary is pressed against the powder to insert the fine dry powder into it. This capillary tube is held between two fingers, gently but firmly, and tapped on the working bench when the sample gradually reaches the sealed end. A little practice will make the operation quite easy. Alternately, the capillary tube can be dropped through a long glass tube held vertically and the sealed end of the capillary downwards. As soon as the sealed end strikes against the floor, the sample is forced to enter the capillary to reach the sealed end. But the former method is more simple and widly practised.

Now this capillary is pressed against the surface of the thermometer and slowly pushed downward until the sealed end and the bulb of the thermometer are on the same level. The capillary will stick to the thermometer. Now the melting point bath is slowly heated by a Bunsen burner or a spirit lamp and the gradual changes of the sample contained in the capillary (e.g. change of volume, colour, sharpness of the m.p, or wide range of m.p., decomposition of the sample at the m.p., gas evolution at the m.p, etc) are noted.

When the m.p. of the sample is high (>225°C) then the m.p. is reported within the range of 2°– or 3°C (e.g. m.p. 245–247°C; not like 245–47°) instead of sharp m.p. (e.g. m.p. 245°C). If the sample decomposes at the m.p. then after the m.p, (d) is written [e.g. m.p. 265–268(d)].

Apparatus fitting for distillation of a liquid sample varies on the nature of the distillate and its b.p. (if very high it is to be distilled under vacuum or low pressure induced by water aspirator), collection of different fractions during distillation involving complicated fittings. So it is very desirable that initially the supervisors should show the learners.

It is to be noted that distillation should never be carried out in a closed system and without the boilizers. The b.p. is dependent on the atmospheric pressure and altitude. Distillation under reduced pressure requires capillary tube reaching upto the bottom of the distillation flask which helps bubbling of air through the liquid sample. Impurities in the liquid sample elevates the b.p. while m.p. is depressed by impurities present in the sample. The b.p. of the sample will remain constant as long as pure distillate is collected.

(VIII) If the sample is uncommon or prepared in the laboratory then the "observed" b.p./m.p. as well as the "reported" b.p./m.p. (reference of the books or journals, i.e. the source of the information) are to be given. The is called "Reference citation" which includes the Author(s), name of the journal, volume, page number and year of publication.

(IX) After the completion of the identification the L.N.B should be made upto date, and concluded by writing like "From all these experiments it is concluded that the sample was ... . In support of this conclusion the following arguments are put forward".

(X) Preparation of the derivatives name, procedure, collection, m.p. (crude), reported m.p., crystallisation or purification, m.p. of pure derivatives (vide 2.3 H, I).

(XI) Special test(s) of the sample, if any (vide 2.3 k).

(XII) *Final report:*

(a) Name of the sample (IUPAC/common).

(b) Structural formula of the compound (sample).

(c) Any comments on the nature of the sample supplied (like the sample was coloured, possibly due to long standing at room temperature or storage for a long time. It was colourless on distillation and the distillate was used during identification, etc.).

2.6 DISTINCTION BETWEEN SIMILAR PAIRS

(a) Alkane: C_nH_{2n+2} (i) Br_2 soln. not decolourised. (ii) $KMnO_4$ soln. (Baeyer's soln.) not decolourised. (iii) No H_2 intake by catalytic hydrogenation. (iv) Mass spectrometric analysis is helpful to establish identity.	(a) Alkene: C_nH_{2n}. Alkyne: C_nH_{2n-2}. (i) Yellow Br_2 soln. decolourised. (ii) Baeyer's soln. decolourised. Thus: 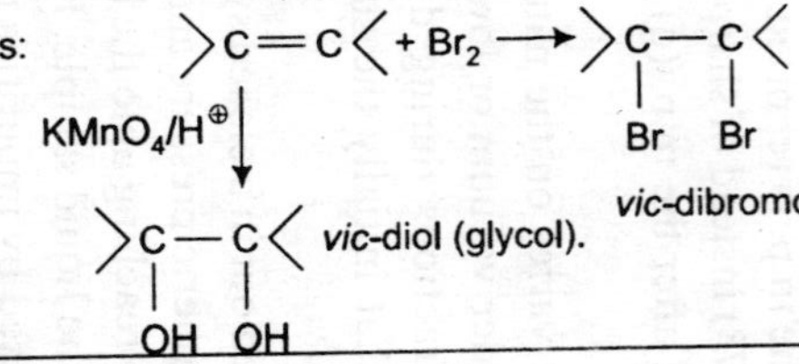
(b) Ethylene, $H_2C=CH_2$ (i) No rxn. with NH_4OH/Cu_2Cl_2 (ii) No rxn. with $NH_4OH/AgNO_3$. (iii) Ozonolysis gives formaldehyde which forms yellow 2,4-DNP.	(b) Acetylene $H–C\equiv C–H$ (i) Red Cu_2C_2 ppt. with NH_4OH/Cu_2Cl_2 (ii) White Ag_2C_2 ppt. with $NH_4OH/AgNO_3$ (iii) No 2,4-DNP of the ozonolysis product.
(c) 1-Butyne, $Et–C\equiv C–H$ (i) Liberates $NH_3\uparrow$ with $NaNH_2$ (Sodamide) (ii) Red ppt. with NH_4OH/Cu_2Cl_2. (iii) White ppt. with $NH_4OH/AgNO_3$. (characteristics of l-yne) (iv) Ozonolysis product (HCO_2H), gives white Ag_2Cl_2 or black Hg when treated with $HgCl_2$.	(c) 2-Butyne, $MeC\equiv CMe$. (i) No $NH_3\uparrow$ with $NaNH_2$ (ii) No rxn. with NH_4OH/Cu_2Cl_2 (iii) No rxn. with $NH_4OH/AgNO_3$ (iv) Ozonolysis product gives Na–Salt which reacts with As_2O_3 to give cacodyl oxide $(Me_2As)_2O$.
(d) Methanol, CH_3OH (i) No rxn. with I_2 + dil. NaOH. Δ. (ii) Oil of wintergreen formed with salicylic acid + H_2SO_4. The ester has characteristic sweet small. (iii) Red hot copper wire converts it into formaldehyde which on treatment with 1% phenyl hydazine hydrochloride + 1% sodium nitroprusside solns + NaOH (dil.) gives deep blue colour.	(d) Ethanol, $CH_3–CH_2–OH$. (i) Yellow CHI_3 with characteristic smell formed with I_2 + NaOH (dil.) Δ. (ii) Sweet smelling ester (less pronounced than oil of wintergreen) is formed with salicylic acid + conc. H_2SO_4, Δ. (iii) Red hot copper wire converts it into acetaldehyde which on treatment with 1% sodium nitroprusside soln. + 1 drop of piperidine ($C_5H_{11}N$) gives deep blue soln.

(e) 1 – Propanol, $EtCH_2 - OH$ (i) No CHI_3 with I_2 + NaOH (dil), Δ (ii) Red hot copper wire forms propanal which responds to Tollens reagent (black Ag powder or Ag–mirror)	(e) 2–Propanol, Me –CH(OH)–Me (i) CHI_3 (yellow solid with smell) is formed with I_2 + NaOH (dil), Δ. (ii) Red hot copper wire forms propanone which does not respond to Tollens reagent.
(f) Formaldehyde, HCHO (Methanal) (i) No CHI_3 with I_2 + NaOH (dil), Δ. (ii) Cannizzaro reaction with conc. NaOH, Δ (but no visible change) (iii) 1% Sodium nitroprusside soln. + 1 drop piperidine, no colour change. (iv) 1% Phenylhydrazine hydrochloride + 1% sodium nitroprusside solns. + NaOH (dil.) gives deep blue soln.	(f) Acetaldehyde, CH_3–CHO (Ethanal) (i) Gives CHI_3 with I_2 + NaOH (dil), Δ (ii) Undergoes polymerisation to give brown ppt. with NaOH, Δ (iii) 1% Sodium nitroprusside + 1 drop piperidine gives deep blue colour. (iv) 1% Phenylhydrazine hydrochloride + 1% Sodium nitroprusside solns. + NaOH (dil) gives no blue soln.
(g) Acetadehyde, CH_3–CHO, (Ethanal). (i) Tollen's reagent: Ag↓ (Black) (ii) Schiff's reagent: Pink colour returns. (iii) Fehling's soln.: red Cu_2O↓. (iv) NaOH (conc), Δ: Brown ppt. (colour) (v) Does not respond to the Legal's test.	(g) Acetone, Me_2CO (2-oxopropane or propanone). (i) Tollen's Reagent: No Ag↓ (ii) Pink colour does not return with Schiff's reagent. (iii) No rxn. with Fehling's soln. (iv) No visible change with NaOH, Δ (v) Aq. soln. of sample (3 ml) + 1% sodium nitroprusside soln. (3 drops) + 1% NaOH soln. (3 drops). makes orange–red colour which turns purple on acidification with HOAC. This is Legal's test for ketones.
(h) Formic acid, HCO_2H, (Methanoic acid) (i) $KMnO_4/H_2SO_4$ dil., Δ; becomes colourless. (ii) Tollens reagent gives black ppt. of Ag↓ (iii) $HgCl_2$ soln. gives white Hg_2Cl_2. with excess of $HgCl_2$ gives black Hg↓. (iv) Does not respond to cacodyl oxide test with $HCO_2Na + As_2O_3$, Δ.	(h) Acetic acid CH_3CO_2H (Ethanoic acid) (i) $KMnO_4/H_2SO_4$ dil, Δ; colour remains unchanged. (ii) Tollens reagent remains unchanged (i.e. no Ag↓). (iii) No reaction with $HgCl_2$ (iv) With CH_3–$CO_2Na + As_2O_3$, Δ, poisonous and obnoxious cacodyl oxide $[(Me_2As)_2O]$ is formed.

(i) Formic acid, HCO_2H (Methanoic acid) (i) $NaHCO_3$ dil. soln. gives $CO_2\uparrow$ bubbles. (ii) No ppt. of 2,4-DNP with 2,4-DNPh. HCl soln. (iii) Moist blue litmus paper turns red. (iv) 1% $C_6H_5NH{\cdot}NH_2{\cdot}HCl$ + 1% $Na_2[Fe(CN)_5NO]$ + NaOH (dil.). no colour change.	(i) Formaldehyde, H_2CO (methanal) (i) Dil. $NaHCO_3$ remains unreacted. (ii) Yellow ppt. of 2,4-DNPh. HCl. (iii) Blue litmus remains unchanged. (iv) 1% Phenyl hydrazine hydrochloride + 1% sodium nitroprusside + dil. NaOH give deep blue soln.
(j) Ethyl acetate, $CH_3–CO_2–C_2H_5$ (Ethyl ethanoate). (i) Not very volatile, b.p. 77°C. (ii) Responds to hydroxamic acid test (S + $NH_2OH{\cdot}HCl$ + X's NaOH, dil, Δ then cooled to r.t., + X's dil. HCl and dil. $FeCl_3$ – blue, violet or brown colour appears. (iii) After hydrolysis the C_2H_5OH formed furnish CHI_3 with I_2 + NaOH (dil.).	(j) Diethyl ether, $(C_2H_5)_2O$.(3-oxapentane). (i) Very volatile, b.p. 34°C. (ii) Does not respond to the hydroxamic acid test of the esters. (iii) No rxn with dil. NaOH, no iodoform.
(k) Ethyl alcohol, C_2H_5OH, (Ethanol) (i) Not very volatile, b.p. 78°C. (ii) With metallic Na, yields H_2 gas. (iii) Reacts with PCl_5 to give HCl gas at room temperature. (iv) I_2 + NaOH (dil.), Δ; yields CHI_3, having yellow colour and characteristic smell, m.p. 119°C. (v) Soluble in H_2O freely in all proportions.	(k) Diethyl ether, Et-O-Et. (i) Very volatile, b.p. 34°C. (ii) No rxn. with metallic Na. (iii) Does not react with PCl_5 at room temperature. (iv) No rxn with I_2 + NaOH (dil.). (v) Insoluble in H_2O; density less than H_2O.
(l) Ethanol, C_2H_5OH (Ethyl alcohol). (i) I_2 + NaOH (dil.), Δ, gives CHI_3 characteristic smell, yellow colour and m.p. 119°C. (ii) Metallic Na liberates H_2 gas. (iii) Does not respond to the hydroxamic acid test.	(l) Ethyl acetate, $CH_3–CO_2–CO_2H_5$ (Ethyl ethanoate). (i) I_2 + NaOH, (dil.), Δ; No CHI_3 formed. (ii) Metallic Na does not liberate H_2. (iii) Responds to the hydroxamic acid test for the esters (vide j (ii)], deep brown soln. results.

(m) Alkanol, R–OH (Alkyl alcohols). (i) $FeCl_3$ soln.; colour remains unchaged. (ii) Ferric–Ferricyanide soln. remains unchanged in colour. (iii) Br_2 water does not react readily. (iv) $K_2Cr_2O_7$ + dil. H_2SO_4, Δ, turns green. The collected rxn. product furnishes 2,4-DNP derivative. (v) Does not respond to the Schotten–Baumann rxn. (vi) Phenolphthalein is not produced.	(m) Phenol, Ar–OH (e.g. C_6H_5OH). (i) $FeCl_3$ soln. colour changed to blue or violet discharged by HOAc or HCl. (ii) Ferric-Ferricyanide colour changes to deep blue due to formation of Prussian blue. (iii) Br_2 water colour discharged with ppt. of bromoderivatives. (iv) No 2,4-DNP derivative after treatment with $K_2Cr_2O_7$ + dil. H_2SO_4, Δ followed by collection of the rxn. pdt. (v) Furnishes benzoate derivatives with Schotten-Baumann rxn. (vi) Phenolphthalein can be made from C_6H_5OH.
(n) Chloroform $CHCl_3$. (i) Combustible, vapour catches fire. (ii) $C_6H_5NH_2$ + dil. NaOH, warmed; generates carbylamine, C_6H_5NC having bad smell.	(n) Carbon tetrachloride, CCl_4. (i) Noncombustible, used as fire extinguisher. (ii) $C_6H_5NH_2$ + dil. NaOH, warmed; carbylamine is not generated.
(o) Aniline sulphate $(C_6H_5NH_2)\ H_2SO_4$ (i) $BaCl_2$ in aq. soln. furnish insoluble $BaSO_4$ ppt. (white). (ii) Does not furnish methyl orange indicator. (p) Aniline hydrochloride, $C_6H_5NH_2{\cdot}HCl$ (i) Aqueous soln. turns blue litmus red, highly soluble in H_2O. (ii) $AgNO_3$ aq. soln. gives white ppt. of AgCl, sol. in NH_4OH and insol. in dil. HNO_3.	(o) Sulphanilic acid, p-HO_3S — 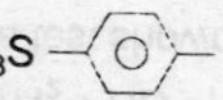 — NH_2 (i) $BaCl_2$, dil. soln does not yield insoluble $BaSO_4$ ppt. (ii) Furnishes methyl orange indicator. (p) Monochloroaniline, $C_6H_4]^{-Cl}_{-NH_2}$ (i) Insoluble in H_2O, does not affect blue litmus. (ii) No ppt. with $AgNO_3$ (dil.).
(q) Benzoyl chloride, C_6H_5–CO–Cl. (i) Aq. soln. of Ag NO_3 yields white AgCl, sol. in NH_4OH and insol. in dil. HNO_3. (ii) Reacts with absolute ethanol to furnish sweet smelling ethyl benzoate, $C_6H_5–CO_2C_2H_5$ at r.t.	(q) Chlorobenzoic acid, $C_6H_4]^{-Cl}_{-CO_2H}$ (i) No rxn. with dil. $AgNO_3$ soln. (ii) No reaction with EtOH at r.t., but reacts with EtOH + Conc. H_2SO_4, Δ to yield ester.

(r) *o*-Phenylenediamine, $C_6H_4]^{-NH_2}_{-NH_2}$ (*ortho*) (i) Reacts with glyoxal, OCH–CHO, to form quinoxalin derivative. m(ii) HNO_2 furnish benzotriazole, C_6H_4 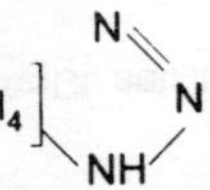 N (*ortho*). No diazo comp.	(r) *m*-Phenylenediamine, $C_6H_4]^{-NH_2}_{-NH_2}$ (*meta*) (i) Quinoxaline deriv. is not formed. (ii) HNO_2 furnish Bismark Brown (a brown compound).
(s) *o*-Phenylenediamine, $C_6H_4]^{-NH_2}_{-NH_2}$ (ortho) (i) Glyoxal reacts to give quinoxalin. (ii) HNO_2 furnishes benzotriazole deriv. (vide r).	(s) *p*-Phenylenediamine, $C_6H_4]^{-NH_2}_{-NH_2}$ (*para*) (i) Quinoxalin deriv. is not formed. (ii) HNO_2 furnishes tetrazo deriv. which couples with alkaline β–naphthol to yield azo-dye.
(t) Ethanol, CH_3–CH_2–OH (i) Responds to iodoform test (ii) Oxidised by hot Cu-wire and the product gives blue colour with 1% Na_2 [Fe(CN)$_5$.(NO)] + piperidine.	(t) Propan–1–ol, CH_3–CH_2–CH_2–OH. (i) Does not yield CHI_3 with I_2 + $HO^{\ominus}$, Δ. (ii) Does not respond to the test shown under Ethanol (t. ii).
(u) Ethanol, CH_3–CH_2–OH (i) I_2 + NH_4OH-warm → No CHI_3. (ii) Does not react with 2,4-DNPh.HCl. (iii) Benzoic acid + Conc. H_2SO_4, Δ, yield ester, sweet smell. (iv) Does not respond to Legal's test	(u) Acetone, CH_3–CO–CH_3 (Propanone). (i) I_2 + NH_4OH, warm gives iodoform (CHI_3). (ii) 2,4-DNP is obtained by 2,4-DNPh HCl. (iii) Does not give ester. (iv) Responds to Legal's test [vide g (v)]
(v) *m*-Nitro aniline, $C_6H_4]^{-NO_2}_{-NH_2}$ (*meta*) (i) Mulliken-Barker test is positive. (ii) HNO_2 does not furnish Bismark-Brown dye.	(v) *m*-Phenylenediamine,-$C_6H_4]^{-NH_2}_{-NH_2}$ (*meta*) (i) Mulliken-Barker test negative. (ii) HNO_2 furnish Bismark-Brown dye.

2.7 LIST OF COMMON DERIVATIVES OF SOME WELL-KNOWN COMPOUNDS

b.p. (m.p.)°C	*Name and formula*	*Derivative I (m.p. °C)*	*Derivative II (m.p. °C)*
Class	*Hydrocarbons, CnHm.*	*Oxidation/Nitration, etc.*	*Picrate/Bromination, etc.*
1. 80	Benzene, C_6H_6	*m*–Dinitrobenzene (90)	
2.110	Toluene, C_6H_5–CH_3	Benzoic acid (122)	Benzaldehyde to 2,4-DNP (237)
3. 144	*o*–Xylene, $C_6H_4(CH_3)_2^{1,2}$	Phtlhalic acid (195)	Phthalic anhydride (133)
4. 146	Styrene, Ph–CH = CH_2	Dibromo (73)	Benzoic acid (122)
5. 151	Cumene, Ph–$CHMe_2$	Benzoic acid (122)	2,4,6-Trinitrocumene (109)
6. 207	Tetralin $C_6H_4(CH_2)_4^{1,2}$	Phthalic acid (195)	5,7-Dinitro deriv. (95)
7. 240	α-Methylnaphthalene $C_{10}H_7$–CH_3 (α)	1-Methyl-4-nitro-naphthalene (71 – 72)	Picrate (141)
8. (70)	Biphenyl Ph–Ph	4,4′-Dinitrobiphenyl (234)	4,4′-Dibromobiphenyle (164)
9. 262	Diphenylmethane Ph–CH_2–Ph	Benzophenone (51)	2, 4-DNP of benzophenone (238).
10. (80)	Naphthalene $C_{10}H_8$	Picrate (150)	1-Nitronaphthalene (60) [HNO_3 at r.t.]
11. (95)	Acenaphthene $C_{10}H_6(CH_2)_2^{1,8}$	Nitro (101)	Picrate (162)
12. (100)	Phenanthrene, $C_{14}H_{10}$	9,10-Quinone (207)	(i) Picrate (143) (ii) Styphnate (142)
13. (114)	Fluonene, $C_{13}H_{10}$	Fluorenone (84) [$KBrO_3$/HOAc]	Picrate (79), *unstab.*
14. (216)	Anthracene. $C_{14}H_{10}$	9,10-Quinone (286)	(i) Picrate (138) (ii) Styphnate (180)
15. –	Alkyl benzenes	Benzoic acid (122).	
16. 83	Cyclohexene, C_6H_{10}	Adipic acid (151).	
Class CAN (+)	Alcohols, R–OH & Ar $(CH_2)_n$–OH.	2,4-DNP from >C = 0 S-one from >C = 0	Amide/Anilide from –COOH.
1. 64	Methanol, MeOH	3,5-Dinitrobenzoate, (109)	2,4-DNP (of CH_2 = 0) (155) [hot CuO]
2. 78	Ethanol, EtOH	Iodoform (119)	2,4-DNP (of CH_3–CHO) (148)
3. 82	Propan-2-ol Me_2CHOH	Iodoform (119)	2, 4-DNP (of Me_2CO) (162)
4. 83	*Tert*-butanol Me_3COH	3,5-Dinitrobenzoate (142)	*p*-Nitrobenzoate (116)
5. 205	Benzyl alcohol Ph–CH_2–OH	Benzoic acid (122)	2,4-DNP (of Ph CHO) (237)
6. 257 (33)	Cinnamyl alcohol Ph–CH = CH–CH_2OH	Benzoic acid (122) [$KMnO_4$]	Cinnamic acid [$K_2Cr_2O_7$] (133) 3,5-Dinitrobenzoate (121)
7. (35)	Pinacol $[Me_2C(OH)]_2$	Iodoform (119)	Tetramethyl ethylene diabromide (192) [Fuming HBr].
8. 219	β-Phenylethanol Ph–CH_2–CH_2–OH	Phenyl acetic acid [76]	3,5-Dinitrobenzoate (108)

Class	Phenols	Deriv. I (m.p °C) miscellaneous	Deriv. II (m.p. °C). miscellaneous
1. 181	Phenol, C_6H_5–OH	2,4,6-Tripromophenol (93).	(i) Benzoate (68) (ii) Acetate (197) (iii) Picrate (83)
2. 191	*o*–Cresol $C_6H_4(OH)^1(CH_3)^2$	*p*-Nitrobenzoate (94)	(i) Dibromo cresol (56) (ii) Picrate (87).
3. 201	*m*-cresol.	(i) *p*-Nitrobenzoate (91) (ii) Picrate (88)	Bromo compound (84)
4. 202	*p*-cresol	(i) Benzoate (71) (ii) Acetate (214)	Bromo compound (108)
5. (94)	α-naphthol, $C_{10}H_7(OH)$	(i) Benzoate (56) (ii) Dibromo (105)	Picrate (189)
6. (105)	Catechol, $C_6H_4(OH)_2^{1,2}$	(i) Bromo (192) (ii) Benzoate (84)	Pirate (121)
7. (110)	**Resorcinol** $C_6H_4(OH)_2^{1,3}$	(i) Dibenzoate (117) (ii) Diacetate (278)	Trinitro (175)
8. (122)	β-Naphthol, $C_{10}H_7(OH)^2$	(i) Benzoate (107) (ii) Acetate (71)	Picrate (156)
9. (133)	Pyrogallol, $C_6H_3(OH)_3^{1,2,3}$	Benzoate (89) *tri-*	Acetate (161) *tri-*
10. (169)	Quinol, $C_6H_4(OH)_2^{1,4}$	Bezoate (199) *di-*	Acetate (123) *di-*
11. (218)	Fluoroglucinol $C_6H_3(OH)_3^{1,3,5}$	Benzoate (173) *tri-*	Acetate (105)

Class	Aldehydes –CHO	2,4,-DNP	(i) S-one (ii) –COOH
1. – 21	Methanal, HCHO	2,4-DNP (162)	S-one (169)
2. 21	Ethanal, CH_3–CHO	2,4-DNP (148)	S-one (163) (ii) CHI_3 (119)
3. 161	Furfural, $C_4H_3O^1(CHO)^2$	2,4-DNP (207)	S-one (201) unstab.
4. 179	Benzaldehyede C_6H_5–CHO	2,4-DNP (237)	(i) S-one (167). (ii) Benzoic acid (122)
5. 194	Phenylacetaldehyde Ph–CH_2–CHO	2,4-DNP (261)	(i) S-one (231). (ii) Phenylacetic acid (76)
6. 197	Salicylaldehyde $C_6H_4(OH)^2(CHO)^1$	(i) 2,4-DNP (151). (ii) S-one (231).	(i) Salicylic acid (157).
7. 221	Cinnamaldehyde Ph–CH = CH–CHO	(i) 2,4-DNP(227) (ii) S-one (215).	(i) Cinnamic acid (133) (ii) Benzoic acid (122)
8. (106)	*m*–Hydroxybenzaldehyde, $C_6H_4(OH)^3(CHO)^1$	(i) 2,4-DNP (259)	(i) S-One (198)
9. (116)	*p*-Hydroxybenzaldehyde, of $C_6H_4(OH)^4(CHO)^1$	(i) 2,4-DNP (157)	(i) S-one (225).

Class	Ketone $\overset{C}{\underset{C}{}}$>C = O	2,4-DNP.	S-one/other.
1. 56	Acetone, Me_2CO	(i) 2,4-DNP (128)	(i) S-one (187) (ii) CHI_3 (119)
2. 80	Methylethylketone, (MEK). MeCOEt	2,4-DNP (114)	S-one (140)
3. 102	Diethyl ketone (DEK), Et_2CO	2,4-DNP (154)	S-one (139)
4. 139	Acetyl acetone MeCO–CH_2–COMe	2,4-DNP (209)	(i) Dimethylpyrazolecarboxylic imide, (109) (during S-one preparation).
5. 192	Acetonylacetone $(MeCO–CH_2)_2$	2,4-DNP (257)	S-one (221)
6. 130	Mesityl oxide Me_2C = CH–COMe	2,4-DNP (203)	(i) S-one (157) (ii) CHI_3 (119)
7. 156	Cyclohexanone $(CH_2)_5$ CO	(i) 2,4-DNP (161)	(i) S-one (166) (ii) Adipic acid (151).
8. 202	Acetophenone Ph–CO–Me	(i) 2,4-DNP (250) (ii) CHI_3 (119)	(i) S-one (199) (ii) Benzoic acid (122)
9. 224	*p*-Methylacetophenone $C_6H_4(Me)^4(COMe)^1$	2,4-DNP (258)	(i) S-one (204) (ii) CHI_3 (119)
10. (52)	β-Acetylnaphthalene $C_{10}H_7$–CO–Me (β)	2,4-DNP (261)	(i) S-one (234) (ii) CHI_3 (119)
11. (42)	Benzal acetone Ph–CH = CH–COMe	2,4-DNP (227)	S-one (185)
12. (51/49)	Benzophenone Ph_2CO	2,4-DNP (238)	(i) S-one (140) (ii) Benzoic acid (122) (KOH treatment).
13. (95)	Benzil $(Ph\ CO)_2$	2,4-DNP (189)	(i) Benzoic acid (122) (ii) Benzilic acid (151)
14. (106)	Pinacolone, $MeCOCMe_3$	(i) 2,4-DNP (125) (ii) CHI_3 (119)	S-one (157)
15. 166	Diacetone alcohol $Me_2C(OH)$, CH_2–COMe	2,4-DNP (202)	Iodoform (119)
16. (116)	*p*-Benzoquinone $C_6H_4O_2$ (*p*)	Picrate (179)	S-one [243(d)]
17. (130)	Benzoin, Ph–CO–CH (OH)–Ph	(i) 2,4-DNP (245) (ii) Benzoic acid (122)	(i) S-one (205) (ii) Benzil (95)
18. (208)	9,10-Phenanthrene-quinone $C_{14}H_8O_2$ (9, 10)	9-Carboxy-9-hydroxy-fluorene (164)	S-one (221)
19. (109)	*p*-Hydroxyacetophenone, $C_6H_4(OH)^4$ $(COMe)^1$	2,4-DNP (261)	*p*-Hydroxybenzoic (213)

b.p./(m.p.) °C	Name and Formula	Deriv. I (m.p.°C)	Deriv.II (m.p.°C).
Class	Carboxylic acid–COOH	Amide/other	Anilide/other
1. 118	Acetic acid, $MeCO_2H$	Amide (82)	(i) Acetanilide (114) (ii) *p*-Toluidide (147)
2. 141	Propionic acid C_5H_5–CO_2H	Amide (79)	(i) Anilide (102) (ii) *p*-Toluidide (124)
3. 15	Methylpropanoic acid MeCH (Me)–CO_2H	Amide (129)	Anilide (105)
4. 163	*n*-Butyric acid Me (CH_2)–CO_2H	Amide (115)	Anilide (90)*
5. (18)	Lactic acid MeCH (OH) –CO_2H	Amide (79)	Anilide (58)
6. (76)	Phenylacetic acid Ph–CH_2–COOH	Amide (154)	(i) Anilide (117) (ii) SBT (165)
7. (97)	Glutaric acid $(CH_2)_3$ $(CO_2H)_2$	Amide (174)	(i) Anilide (224) (ii) *p*-Toluidide (218)
8. (100)	Citric acid $(HO_2C$–$CH_2)_2C(OH)\cdot$ (CO_2H)	Amide (215)	(i) Anilide (199) (ii) *p*-Toluidide (189)
9. (101)	Oxalic acid $(CO_2H)_2\cdot$ $2H_2O$	*p*-Toluidide (267)	Anilide (245)
10 (102)	*o*-Toluic acid $C_6H_4(CH_3)^2(CO_2H)^1$	Amide (142)	(i) Anilide (125) (ii) SBT (146)
11. (101)	Malic acid HO_2C–CH· (OH)–CH_2–CO_2H	Amide (158)	Anilide (197)
12. (111)	*m*-Toluic and C_6H_4 $(Me)^3$ $(CO_2H)^1$	Amide (94)	Anilide (126)
13. (118)	*dl*—Mandelic acid Ph–CH(OH)–CO_2H	Amide (134)	Anilide (151)
14. (122)	Benzoic acid Ph–CO_2H	(i) Amide (128) (ii) *m*-Nitrobenzoic acid (140)	(i) Anilide (161) (ii) *p*-Toluidide (158)
15. (133)	Cinnamic acid Ph–CH = CH–CO_2H	(i) Amide (141) (ii) Dibromo (195) (iii) *p*–Nitrocinnamic acid (287) [Conc. HNO_3]	(i) Anilide (153) (ii) *p*-Toluidide (168)
16. (133)	Malonic acid $CH_2(CO_2H)_2$	(i) Amide (170)	Tribromoacetic acid (136)
17. (133)	Furoic acid $C_4H_3O^1(CO_2H)^2$	Amide (141)	Anilide (123)

b.p/(m.p °C)	*Name & Formula*	*Deriv.I. (m.p. °C)*	*Deriv II (m.p. °C).*
Class	*Carboxylic acids –COOH*	*Amide/other*	*Anilide/other*
18. (130)	Maleic acid $C_2H_4\ (COOH)_2$ [*cis*].	(i) Amide (181) (ii) Anhydride (56)	(i) Anilide (167) (ii) *p*-Toluidide (142)
19. (137)	*o*-Chlorobenzoic acid $C_6H_4Cl^2\ (CO_2H)^1$	(i) Amide (139) [$NaHCO_3$ insoluble]	Anilide (114)
20. (148)	β-Phenylpropionic acid Ph–CH_2–CH_2–COOH	Amide (105)	(i) Anilide (98) (ii) *p*-Toluidide (135)
21. (150)	Benzilic acid $Ph_2C(OH)–CO_2H$	Amide (154)	(i) Anilide (175) (ii) Benzophenone (51)
22. (151)	Adipic acid $HO_2C(CH_2)_4$–COOH	(i) Amide (220) (ii) 2,4-DNP of cyclopentanone (146)	(i) Anilide (235) (ii) S-one (206) of cyclopentanone
23. (157)	Salicylic acid $C_6H_4\ (OH)^2\ (CO_2H)^1$	(i) Amide (139) (ii) Aspirin (135)	(i) Anilide (134) (ii) 5-Nitrosalicylic acid (230) [15% HNO_3, Δ]
24. (161)	α-Naphthoic acid $C_{10}H_7\ (COOH)^1$	(i) Amide (202)	(i) Anilide (163)
25. (169)	*dl*-Tartaric acid $[HO_2C–CH–(OH)]_2$	Amide (196)	Anilide (264)
26. (184)	*p*-Anisic acid $C_6H_4\ (OMe)^4\ (CO_2H)^1$	(i) Amide (162) (ii) *p*-Hydroxybenzoic acid (213) [48% HBr, Δ]	Anilide (168).
27. (185)	Succinic acid $(CH_2)\ (CO_2H)_2$	(i) Anhydride (120) (ii) Amide (242)	(i) Anilide (226) (ii) *p*-Toluidide (220)
28. (185)	β-Naphthoic acid $C_{10}H_7\ (COOH)^2$	Amide (192)	Anilide (170)
29. (178)	*p*-Toluic acid $C_6H_4\ (OH)^4\ (COOH)^1$	Amide (159)	Anilide (140)
30. (189)	Oxalic acid $(CO_2H)_2$	Urea oxalate (171)	
31. (195)	Phthalic acid $C_6H_4\ (CO_2H)_2^{1,2}$	Anhydride (130)	(i) Phthalimide (238) (ii) *p*-Nitrobenzoate (153)
32. [200 (d)]	Coumaric acid $C_6H_4\ (OH)^2$. $(CH = CH{\cdot}CO_2H)^1$	Acetyl (146)	Salicylic acid (157) [KOH fused]
33. (201)	*m*-Hydroxybenzoic acid $C_6H_4(OH)^3(CO_2H)^1$	(i) Amide (168) (ii) Acetyl (127)	Amilide (157)
34. (213)	*p*-Hydroxybenzoic $C_6H_4\ (OH)^4\ (CO_2H)^1$	(i) Amide (161) (ii) Acetyl (185)	Anilide (197)
35. (213)	Resorcilic acid $C_6H_4\ (OH)_2^{2,4}\ (CO_2H)^1$	Amide (220)	Anilide (127)
36. (287)	Fumaric acid $HO_2CCH = CH–CO_2H$ [*trans*]	Amide (266)	Anilide (314)

Class	Halo-Compounds	Deriv. I. (m.p.°C)	Deriv.II. (m.p.°C)
1. 185	*p*-Bromotoluene $C_6H_4(Br)^4(CH_3)^1$	*p*-Bromobenzoic acid (251)	Nitro derivative (47)
2. (50)	Bromoacetic acid $BrCH_2–CO_2H$	Amide (91)	Anilide (130)
3. (55)	Trichloroacetic acid, $Cl_3C–CO_2H$	Amide (141)	Anilide (94)
4. (63)	*p*-Bromophenol $C_6H_4(Br)^4(OH)^1$	Benzoate (102)	3,5-Dinitrobenzoate (191)
5. (63)	Chloroacetic and $Cl–CH_2–COOH$	Amide (119)	Anilide (134)
6. (87)	2,4,6-Tribromoanisole $C_6H_2(Br_3)^{2,4,6}(OMe)^1$	2,4,6-Tribromophenol (95)	
7. (119)	Indoform CHI_3	Picrate (143)	—
8. (137)	*o*-Chlorobenzoic acid $C_6H_4(Cl)^2(CO_2H)^1$	Amide (142)	Anilide (114)
9. (236)	*p*-Chlorobenzoic acid $C_6H_4(Cl)^4(CO_2H)^1$	Amide (179)	Anilide (194)

Class	Nitro-carboxylic acid	Deriv. I. (m.p.°C)	Deriv. II. (m.p.°C)
1. (141)	*m*-Nitrobenzoic acid $C_6H_3(NO_2)^3(CO_2H)^1$	Amide (142) ($NaHCO_3$—insoluble)	Anilide (153)
2. (144)	*o*-Nitrobenzoic acid, $C_6H_4(NO_2)^2(CO_2H)^1$	Amide (174)	Anilide (155)
3. (205)	*m*-Nitrocinnamic acid $C_6H_4(NO_2)^3$ $(CH = COCO_2H)^1$	Amide (196)	*m*-Nitrobenzoic acid (141)
4. (230)	5-Nitrosalicylic acid $C_6H_3(OH)^2(NO_2)^5$. $(CO_2H)^1$	Amide (225)	Anilide (223)
5. (237)	*o*-Nitrocinnamic acid, $C_6H_4(NO_2)^2$. $(CH = CH\text{-}CO_2H)^1$	Amide (184)	*o*-Nitrobenzoic acid (144)
6. (241)	*p*-Nitrobenzoic acid $C_6H_4(NO_2)^4(CO_2H)^1$	Amide (200)	Anilide (204)

Class	Amide and Anilide formula	Deriv. I Carboxylic acids/other	Deriv. III other
1. (82)	Acetamide $MeCONH_2$	Picrate (107)	Acetanilide 114) [$PhNH_2$, Δ].
2. (114)	Acetanilide Ph–NH–COMe	Benzanilide (161) [*via* hydrolysis]	*p*-Bromoacetanitide (167) [Br_2/HOAc]
3. (125)	Succinimide $(CH_2–CO)_2NH$	Succinic acid, (185)	Succinanil (156) [$PhNH_2$, Δ]
4. (128)	Benzamide $Ph–CONH_2$	Benzoic acid (122)	(i) Aniline to acetyl (114) (ii) Aniline to tribrome aniline (119) and (iii) Benzanilide (161).

5. (132)	Urea $OC(NH_2)_2$	Nitrate (163) [sat. soln. + conc. HNO_3]	Oxalate (171) [sat. soln. + sat. oxalic acid].
6. (139)	Salicylamide $C_6H_4(OH)^2$ $(CONH_2)^1$	Salicylic acid (157)	—
7. (161)	Benzanilide C_6H_5–CO–NHPh	Benzoic acid (122)	*p*-Bromobenzanilide (198)

Class	Nitro-Compds $-NO_2$	Deriv. I. Misc.	Deriv. Misc.
1. 126	Tetranitromethane $C(NO_2)_4$(TNM)	Colour Test	—
2. 211	Nitrobenzene Ph–NO_2	(i) Benzanilide [161]	*m*-Dinitrobenzene (90)
3. 222	*o*-Nitrotoluene $Ph(NO_2)^2(CH_3)^1$	*o*-Nitrobenzoic acid (144)	—
4. 265	*o*-Nirtroanisole $C_6H_4(NO_2)^2$ $(OMe)^1$	—	2,4-Dinitroanisole (94)
5. (54)	*p*-Nitroanisole $C_6H_4(NO_2)^4(OMe)^1$	—	*p*-Nitrophenol (114)
6. (79)	β-Nitronaphthalein $C_{10}H_7(NO_2)$–β	4-Nitrophthalic acid (164) [CrO_3/HOAe]	—
7. (90)	*m*-Dinitrobenzene C_6H_4 $(NO_2)_2{}^{1,3}$	*m*-Nitroaniline (112) [NH_4SH/EtOH]	—

Class	*Nitro. alcohols (or phenols) or Nitro oxo*	*Deriv. I [O/H]*	*Deriv. II [Substitution]*
1. (44)	*o*-Nitrobenzaldehyde $C_6H_4(NO_2)^2$ $(CHO)^1$	*o*-Nitrobenzoic acid (144)	(i) 2,4-DNP (192) [Vogel. m.p. 256]
2. (74)	*o*-Nitrobenzylalcohol $C_6H_4(NO_2)^2$ $(CH_2OH)^1$	*o*-Nitrobenzoic acid (144)	—
3. (81)	*m*-Nitroacetophenone $C_6H_4(NO_2)^3(COMe)^1$	*m*-Nitrobenzoic acid (140)	(i) 2,4-DNP (227) (ii) CHI_3 (119).
4. (97)	*m*-Nitrophenol $C_6H_4(NO_2)$ $(OH)^1$	Bromo compd. (91)	Benzoate (95).
5. (106)	*p*-Nitrobenzaldehyde C_6H_4 $(NO_2)^4$ $(CHO)^1$	*p*–Nitrobenzoic acid (239)	(i) 2,4-DNP (319) (ii) Oxime (132)
6. (114)	*p*-Nitrophenol C_6H_4 $(NO_2)^4$ $(OH)^1$	Bromo compd. (142)	(i) Benzoate (142) (ii) Acetate (81)
7. (193)	*p*-Nitrobenzylalcohol $C_6H_4(NO_2)^4$ $(CH_2OH)^1$	*p*-Nitrobenzoic acid (239)	—

Class	*Amines* $>\ddot{N}-$	*Benzoyl/other*	*Acetyl/other*
1. 183	Aniline, Ph–NH_2	Benzoyl (161)	(i) Acetyl (114) (ii) Tribromoaniline (118)
2. 197	*o*-Toluidine, $Ph(Me)^2$ $(NH_2)^1$	Benzoyl (143)	(i) Acetyl (112) (ii) Picrate (213)
3. 199	*m*-Toluidine, $Ph(Me)^3$ $(NH_2)^1$	Benzoyl (125)	(i) Acetyl (65) (ii) Picrate (213)

4. (45)	*p*-Toluidine Ph(Me)^4 (NH$_2$)^1	Benzoyl (158)	(i) Acetyl (148) (ii) Picrate (181)
5. (71)	*o*-Nitroaniline $C_6H_4(NO_2)^2(NH_2)^1$	Benzoyl (94)	(i) Acetyl (92) (ii) Picrate (72)
6. (102)	*o*-Pheylenediamine $C_6H_4\ (NH_2)_2^{1,2}$	Dibenzoyl (301)	(i) Diacetyl (185) (ii) Picrate (195)
7. (114)	*m*-Nitroaniline $C_6H_4(NO_2)^3(NH_2)^1$	Benzoyl (157)	Acetyl (155)
8. (122)	*m*-Aminophenol $C_6H_4(NO_2)^3(OH)^1$	Benzoyl (153)	Diacetyl (101)
9. (126)	Benzedine p,p^1–$H_2N–(C_6H_4)–NH_2$	Dibenzoyl (352)	Diacetyl (317)
10. (147)	*p*-Nitroaniline $C_6H_4(NO_2)\ (NH_2)$	Benzoyl (199)	Acetyl (215)
11. (174)	*o*-Aminophenol $C_6H_4(NH_2)^2(OH)^1$		(i) Monoacetyl (201) (ii) Diacetyl (124)
12. (176)	2,4-Dinitroaniline $C_6H_3\ (NO_2)^{2,4}\ (NH_2)^1$	Benzoyl (220)	Acetyl (120)
13. (184)	*p*-Aminophenol $C_6H_4\ (NH_2)^4\ (OH)^1$	Dibenzoyl (234)	Diacetyl (150)
14. (246)	Carbazole	(i) Benzoyl (98) (ii) Acetyl (69)	Picrate (182)
15. 115.	Pyridine, C_5H_5N	Methoiodide (117)	Picrate (167)

2.8 SPECIAL TESTS FOR SOME COMPOUNDS (ON THE BASIS OF COLOUR AND ODOUR)

1. *Acetaldehyde:* (i) In aqueous solution $\xrightarrow[\text{hydrochloride}]{\text{dil. benzidine}}$ yellow colouration.

   (i) In aqueous solution $\xrightarrow[\text{+ piperidine(1 drop)}]{5\%\ Na_2[Fe(CN)_5NO]\ (0.5\ ml)}$ deep blue colour.

2. *Acetone*: Sample + *o*–Nitrobenzaldehyde + dil. NaOH → ppt.

3. *Adipic acid:* (i) Sample + Resorcinol + Conc. H_2SO_4, Δ $\xrightarrow[\text{dil. NaOH}]{\text{followed by}}$ purple fluorescence.

   (ii) Sample + $FeCl_3$ solution → Violet.

4. *Anilide* $\xrightarrow[\text{5 min.}]{H_3O^{\oplus},\ \Delta}$ Aromatic 1° amine $\xrightarrow[\Delta]{\text{(i) } CHCl_3\text{, dil. NaOH}}$ (Foul smelling and poisonous carbylamines).

   Aromatic 1° amine ↓ $\xrightarrow[\text{(iii) (dil.NaH + β–naphuthol)}]{\text{(ii) } HNO_2 \text{ (ice cold)}}$ brilliant red Azo-dye.

5. *Aniline:* (vide 4.)

6. Anthracene (and also *phenanthrene*):

(i) When sprinkled on a Bunsen flame it burns with scintillation like shining stars.

(ii) Sample + anhydrous $AlCl_3$ (warmed) $\xrightarrow[\text{t.t.}]{\text{in}}$ violet colour (with napthalene blue colour) developed.

7. *Benzaldehyde:* $\xrightarrow[\text{+ conc. } H_2SO_4\text{, }100°C]{\text{Sample + dimethylaniline (2 parts)}}$ Malachite green dye.

8. Benzidine: Sample/CS_2 soln. $\xrightarrow[\text{(dropwise)}]{\text{aqueous } Br_2}$ Aq. soln. green and gradually became colourless but the CS_2 layer was red.

9. *Benzophenone:* Sample $\xrightarrow[\textit{Molten}]{\textit{Metallic Na}}$ deep blue colour.

10. *Catechol:* Reduces Fehling's solution and Tollens Reagent.

11. *Acetic acid:* Sample $\xrightarrow[\textit{into}]{\textit{Converted}}$ $CH_3–CO_2Na$ $\xrightarrow[\Delta]{AS_2O_3}$ Foul smelling and poisonous cacodil oxide is formed.

12. *Chloroform:* Sample $\xrightarrow[\text{dil. NaOH, }\Delta]{\text{+ Aniline then}}$ Foul smelling poisonous carbylamine is formed.

13. *Citric acid:* In aqueous solution $\xrightarrow[\text{Reagent}]{\text{Deniges}}$ Reaction Mixture aq. $KMnO_4$ soln. (dropwies) violet colour is destroyed with appearance of white ppt.

14. *2,4-Dinitrotoluene:* Sample $\xrightarrow[\text{+ KOH, }\Delta]{\text{Acetone}}$ blue $\xrightarrow{\text{HOAc}}$ violet colour.

15. *m-Dinitrobenzene:* Sample $\xrightarrow{SnCl_2 \text{ soln.}}$ violet colourartion.
$\quad\quad\quad\quad\quad\quad\quad\quad\quad$ $\xrightarrow[\text{glucose}]{}$ – do –

16. *Ethanol*: Sample $\xrightarrow{[O], CuO}$ Acetaldehyde $\longrightarrow$ (vide Test 1).

17. *Formic acid*: Sample $\xrightarrow{HgCl_2 \text{ aq. soln.}}$ white (Hg_2Cl_2) or black (Hg) ppt.

18. *Furoic acid*: Sample + NH_4OH soln. $\xrightarrow[\text{conc. HCl held on the vapour}]{\text{1) }\Delta\text{, Vapour on a Stick moistened with}}$ wood stick became red.

19. *Glycerol*: Sample $\xrightarrow{KHSO_4\text{, }\Delta}$ Acrolein (extremely pungent smell)

20. *Lactic acid*: Sample $\xrightarrow[\text{+5\% ethanolic guaicol}]{\text{conc. } H_2SO_4}$ deep red colouration.

21. *Malic acid*: Sample $\xrightarrow[\text{+conc. } H_2SO_4]{\beta\text{−naphthol}}$ Greenish-yellow fluorescence.

22. *Malonic acid*: Sample $\xrightarrow{Ae_2O, \Delta}$ Orange-green fluorescence.

23. *Methanal*: Sample $\xrightarrow[\text{(ii) } K_3Fe(CN)_6 \text{ (few drops)}]{\text{(i) 1\% Ph-NH MH}_2\cdot\text{HCl}}$ R.M. $\xrightarrow{HCl}$ bright red colouration.

24. *Methanol:*

(i) Sample $\xrightarrow{[O]}$ Methanal $\xrightarrow[\text{above}]{\text{Test 23}}$ bright red colour (Schryver's Test)

(ii) Dry sample $\xrightarrow[\text{(b) Poured on cold water}]{\text{(a) Salieylic acid + conc. } H_2SO_4, \Delta}$ Characteristic smell of oil of wintergreen.

25. *d-Mannitol*: Small amount of sample $\xrightarrow[CuCl_2 \text{ soln.}]{\text{ammoniacal}}$ blue ppt.

26. *Mesityl oxide*: $\xrightarrow{\text{sample} + I_2 + \text{dil. NaOH}}$ yellow ppt. of CHI_3 (smell).

27. *Mucic acid*: (Responds to 18, for furoic acid).

28. *α-Naphthol (or β-Naphthol)*: Sample $\xrightarrow[50°C]{CHCl_3/KOH \text{ soln.}}$ blue colouration

29. *o-Nitrobenzaldehyde*: $\xrightarrow[+NaOH]{\text{Sample + Acetone}}$ blue ppt.

30. *o-Nitrophenol*: Sample $\xrightarrow[\text{warmed}]{\text{NaOH soln.}}$ red colouration (NB: all nitrophenols develop yellow or brown or red colour).

31. *Oxalic acid*: Sample $\xrightarrow[\text{soln.}]{NaHCO_3}$ Na oxalate soln. $\xrightarrow[\text{soln.}]{CaCl_2}$ white ppt.

32. *m-Phenylenediamine*: Sample $\xrightarrow[\text{ii) 1\% MeCHO in EtOH}]{\text{i) dil. HCl}, \Delta}$ yellow-green fluorescence.

33. *Phthalic acid*: Sample $\xrightarrow[\text{ii) conc. } H_2SO_4, \Delta]{\text{i) Resorcinol}}$ R.M. $\xrightarrow{\text{dil.NaOH}}$ greenish–brown fluorescence.

34. *Phthalic anhydride/Phalimide/Diethyl Phthalate* → vide No, 33 test.

35. *Phenol:* Sample (1 Part) + Phthalic anhydride (2 parts) $\xrightarrow[\Delta]{\text{conc. } H_2SO_4}$ R.M. $\xrightarrow[\text{excess}]{NaOH}$ red colour. (N.B.-Product is phenolphthalein indicator)

36. *Resorcinol:* vide No. 33 Test.

37. *Salicylic acid:* Sample + pure MeOH $\xrightarrow[\Delta]{\text{conc. } H_2SO_4}$ on cold water (sweet smell of oil of wintergreen)

38. *Succinic acid:* Sample + Zn + NH_4OH $\xrightarrow[\text{ii) Stick (moist) with conc. HCl}]{\text{i) }\Delta\text{, vapour}}$ stick becomes red.

39. *Sulphanilic acid:* $\xrightarrow[\text{ii) Dimethylaniline in dil. Hcl}]{\text{i) dil. cold } HNO_2}$ Methyl orange indicator (orange in dil. alkali red in dil. acid). (N.B. HOAc may replace dil. HCl)

40. *Tartaric acid:* Sample $\xrightarrow[\text{ii) aqueous } KMnO_4]{\text{i) Deniges reagent, }\Delta}$ purple colour of $KMnO_4$ is discharged but no ppt. (cf citric acid, 13) $\xrightarrow{\text{. }\beta\text{–Naphthol + conc. } H_2SO_4}$ Green colour $\xrightarrow{\Delta}$ No change.

41. *Tertranitromethane (TNM)* $\xrightarrow[\text{ethanol so ln.}]{\text{Pyridine in}}$ unchanged.

42. *Toluidine (o and m):* Sample $\xrightarrow[\text{powder}]{\text{bleaching}}$ violet, yellow-brown or brown colouration. (NB. *p*-Toluidine is exceptional, it gives only colourless mixture).

43. *Urea* $\xrightarrow{\Delta\text{, 150°C}}$ Biurate $\xrightarrow[\text{aq. so ln.}]{CuSO_4}$ blue-violet ppt.

NOTES

- These special tests could be done in t.t. taking small quantities of the starting material and the reagents.
- In these special tests the starting materials and the organic reactants are complementary to each other, that means if the reactant is known the starting material will be unknown and *vice versa.*

2.9 ULTRAVIOLET (U.V.) SPECTROSCOPY

Introduction: A very simple and useful technique for structure elucidation of organic molecules is u.v. spectroscopy which involves electronic excitation and expressed as λ (wavelength) nm of usual range between 175 and 350 nm (approx.); presentation of data is sometimes accompanied by intensity of absorption (ε, epsilon) as shown below.

Examples:

(a) Isolated C = C (acyclic or alicyclic) absorbs between 175 and 200 nm depending on the number of alkyl substitutions. More alkyl groups have the bathochromic effect on the λ_{max}.

Calculation of : λ_{max}	$\lambda_{max, nm}$
(i) Parent open chain/heteroannular diene (conjugated)	= 214(e.g. A).
(ii) Parent homoannular diene (conjugated)	= 253(e.g. B).
(iii) For each extra C=C conjugated with (i) or (ii) (Add)	= 30(e.g. C).
(iv) For each alkyl gr./ring residue on C=C, (Add)	= 5(e.g. D).
(v) Exocyclic C=C (coming out of any ring), (Add)	= 5(e.g.D).

(vi) For OAc gr., OR, SR, Cl/Br substitution (Add) = 0, 6, 30 and 5, respectively.

Notes: (i) Abnormal molecular overcrowding or strain leads to abnormal λ_{max} value (e.g. E).

(ii) λ_{max} of alkenes (conjugated or isolated) are independent of solvents.

(iii) For conjugated alkenes the ε values are between 6,000 and 35,000 (usually).

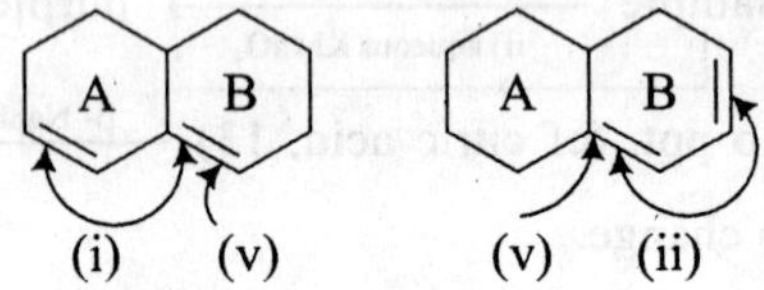

[A] [B]

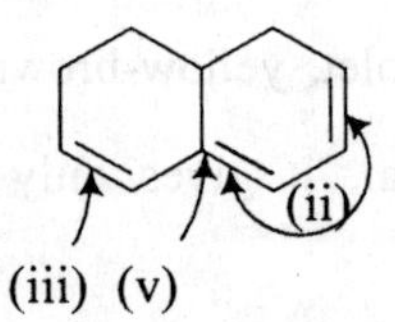

[C]

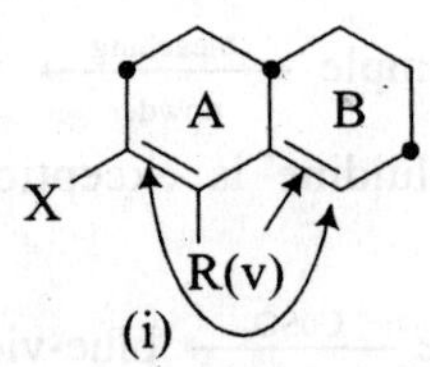

[D]

X = Cl or Br
R = alkyl gr.
• = ring residues
(v) = exo of ring A.
(i) = heteroannular diene.

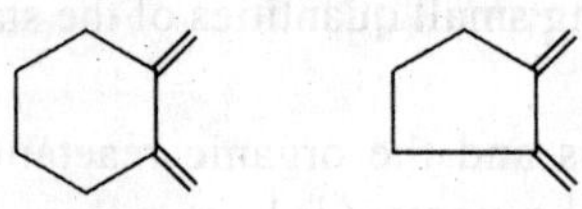

- 5-membered ring compound is almost planar, hence of negligible strain, cf..E

[E]	[E_1]
λ_{obs} 220 nm	λ_{obs} 248 nm
(ε 5.5×10^3)	(ε 15.5×10^3)

[F]

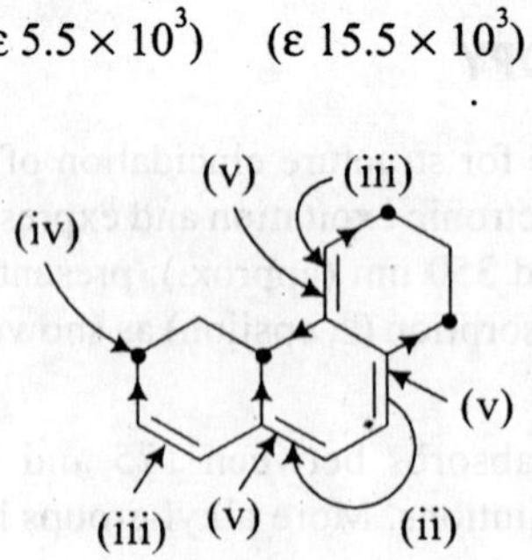

(1) Priority of (ii) > (i) (i.e. heteroannular < homoannular). Add 253 nm

(2) Ring residue [5 × 5 (each)] (—•), (iv) = Add 25 nm

(3) Extended conjugation [2 × 30 (each)] (iii), = Add 60 nm

(4) Exocyclic double bond [3 × 5 (each)], (v) = Add 15 nm

Total, λ_{max} = 353 nm (calcd)

= 355 nm (obs).

[G] 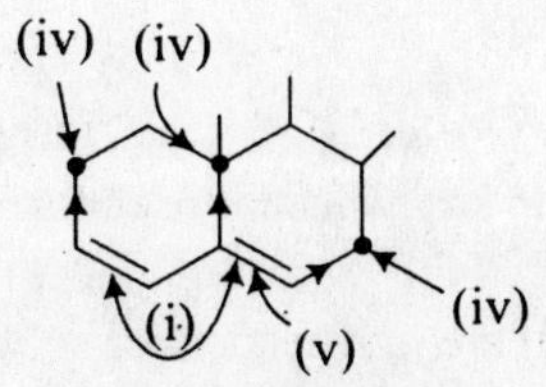

(1) Heteroannular diene, (i) = 214 nm
(2) Ring residue [3 × 5 each) (iv) = 15 nm
(3) Exocyclic double bond (1 × 5) (v) = 5 nm

λ_{max} 234 nm (calcd)

[H] 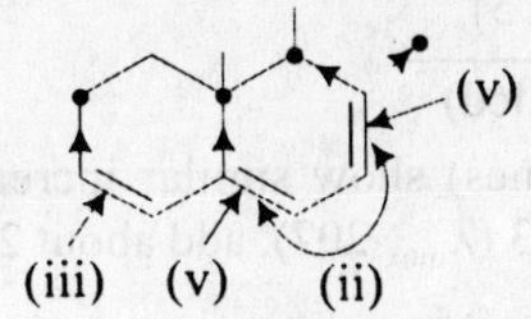

1. Homoannular diene (ii) = 253 nm
2. Ring residue →• (4 × 5) = 20 nm
3. Extended conjugation (1 × 30) (iii) = 30 nm
4. Exocyclic double bond (2 × 5) (v) = 10 nm

Total λ_{max} = 313 nm (Calcd)

($\varepsilon = 19.8 \times 10^3$)

[I] 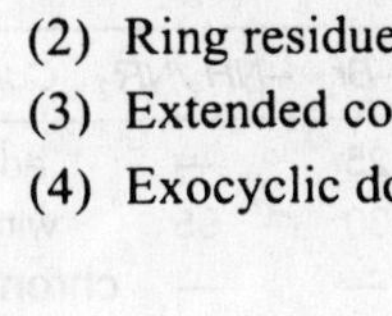

(1) Homoannular diene (ii), = 253 nm
(2) Ring residue → • [3 × 5 (each)] = 15 nm
(3) Extended conjugation (iii) = 30 nm
(4) Exocyclic double bond (v) = 5 nm

Total λ_{max} 303 nm

[J] 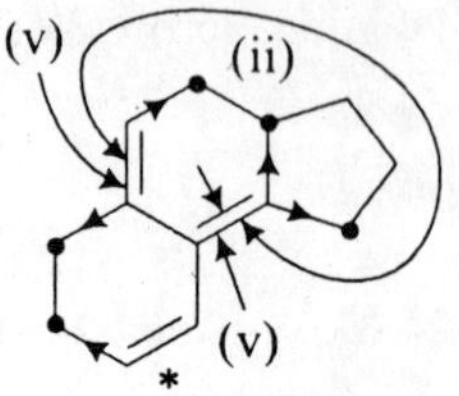

(1) Homoannular diene = 253 nm
(2) Ring residue →• = 25 nm (5 × 5)
(3) Extended conjugation = 0 nm
(not counted)*
(4) Exocyclic C=C =15 nm (3 × 5).

Total λ_{max} 293 (Calcd)

(c) *Conjugated polyynes* (as well as polyenynes) show similar increase in λ. Thus in R (C ≡ C)$_n$ –R (R = Me), when n = 3 (λ_{max} 207), add about 25 nm for every increase of n by one.

(d) *Carbonyl compounds:* (Acetone, λ_{max} 190 (1.86×10^3) and 280 nm (0.01×10^3). Saturated ketones/aldehydes absorb at lower wavelengths and they do not furnish any useful information. However, the α,β-unsaturated oxo compounds with or without substituents are of great utility in structure determination, unlike polyenes these λ_{max} are solvent dependent.

Calculation of λ_{max}

(1) Parent acyclic or cyclohexyl conjugated enone λ_{max} 215 nm.
(2) -Do- cyclopentyl conjugated enone λ_{max} 202 nm.
(3) Parent α,β-unsaturated aldehyde, λ_{max} 207 nm
(4) Each extra C=C conjugated to C=C – C=O add 30 nm
(5) Homoannular diene component add 39 nm
(6) Contribution of substitutions at $C^\delta = C^\gamma - C^\beta = C^\alpha - C^I = O$

Location at	*alkyl/ring residue*	*–OH*	*–OAc*	*–OMe*	*–Cl*	*–Br*	*–NH₂/NR₂*	*Comments.*
α	10 nm	35	6	35	15	25	—	add these
β	12 nm	30	6	30	12	30	95	with parent
γ/δ	18 nm	50	6	γ17/δ31	—	—	—	chromophore

***Note*:** Observed values within ± 5 nm of the calculated values are acceptable.

Examples:

[A] 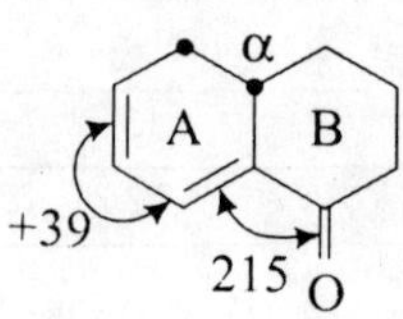

$\lambda_{max}^{\text{Solvent (EtOH)}}$ = 215 + 10 (α,-ring residue •) + 18 (δ-ring residue •) + 5 (exo of ring B) + 30 (extended conjugation) + 39 (homodiene component) = Total 317 nm (Calcd).

[B] 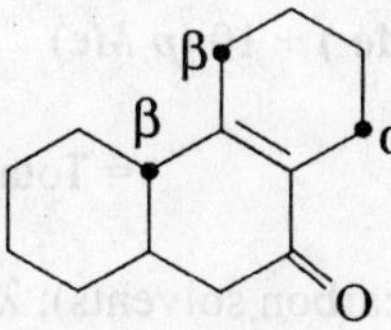 λ_{max}^{EtOH} = 215 + 10 (α) + 24 (β,β) = Total 249 nm

[C] 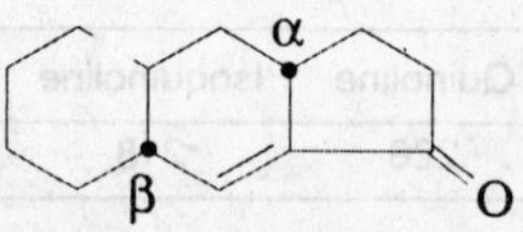 λ_{max}^{EtOH} = 215 + 10 (α) + 12 (β) + 5 (exo) = Total 242 nm

[D] 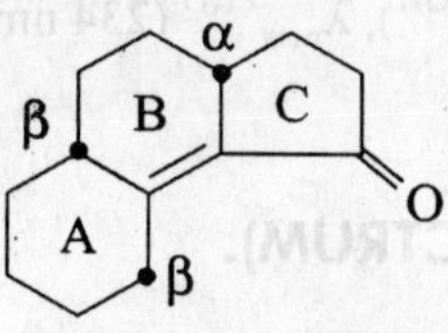 λ_{max}^{EtOH} = 215 + 10 (α) + 24 (β,β) + 5 (exo of A) + 5(exo of C) = Total 259 nm

[E] 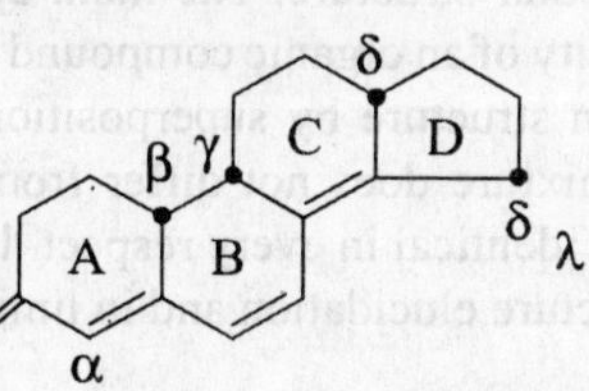 λ_{max}^{EtOH} = 215 + 2(30) (2 extended C=C) + 12 (β) + 3 × 18 (γ/δ) + 5 (exo of B) + 5 (exo of B) + 5 (exo of D) = Total 356 nm.

NOTE: Solvent correction (HEDCEW), add or subtract to reach λ_{max} in EtOH. Hexane/cyclohexane (+ 11), Ether (+ 7), Dioxane (+ 5), Chloroform (+ 1), EtOH/MeOH (0), water (– 8).

(e) *Aromatic compounds:* Benzene 184 (60 × 10^3). Alkyl substitutions on benzene ring has the bathochromic effect. Additional rings fused with parent benzene also have similar bathochromic effect on λ_{max}.

Examples:

(1) Acetophenone (parent compound) λ_{max}^{EtOH} 246 nm.

Substitutions on acetophenone ring (nature and locations),

λ in EtOH solvent in nm.

Location	Alkyl/ring residue	OR¹/OH	Cl	Br	NH_2	NMe_2
o, m	3	7	0	2	13	20
p	10	25	10	15	58	85

(2) 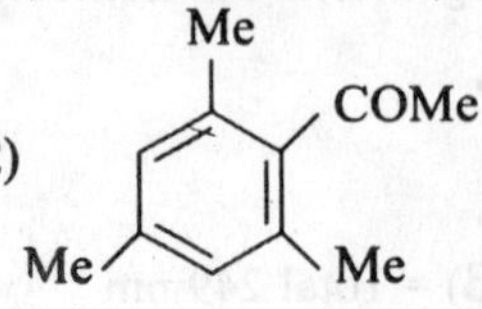 λ_{max} 246 (parent) + 2 × 3 (*o*, *o* Me) + 10 (*p Me*)

= Total 262 nm.

(f) *Heterocyclic compounds* (parent compounds in hydrocarbon solvents); λ_{max} nm.

Compounds	Pyrrole	Thiophene	Pyridine	Quinoline	Isoquinoline	Indole
λ_{max}^{CnHm}	210	231	195	226	218	220

(g) *Example of solvent correction:* If the compound has λ_{max}^{EtOH} 241 nm then the same compound will show λ_{max}^{Hex} (230 nm, add 11 to get λ_{max}^{EtOH}), $\lambda_{max}^{Et_2O}$ (234 nm, add 7 to get λ_{max}^{EOH}), $\lambda_{max}^{H_2O}$(249 nm, subtract 8 to reach λ_{max}^{EtOH}).

2.10 INFRARED SPECTROSCOPY (IR SPECTRUM).

(a) *Infrared spectroscopy (IR spectrum).* Electromagnetic interaction affects the vibrational frequencies of the molecular structure. The main application of the IR spectra is to establish the identity of an organic compound (synthetic or natural) with a compound of known structure by superposition of the two spectra. The structure of recemic mixture does not differ from that of the enantiomers, hence these spectra are identical in every respect. In addition to this the IR spectrum is useful in structure elucidation and in limited cases for quantitative analysis.

(b) The spectrum is taken with the sample in CCl_4 (anhydrous) solvent, in Nujol mull as thin film or in KBr pellets, any one of the procedure is adopted. It should be noted that all of the bands in the spectrum do not require assignments or accounting for.

(c) *Presentation of data:* As frequency, ν_{max} cm^{-1} and/or λ_{max} nm (wavelength). Generally, approximately ν_{max} 4000 cm^{-1} to 1350 cm^{-1} (i.e. λ_{max} 2.5 nm to 7.4 nm; ν_{max} = 10000/λ_{max}) is called the "Characteristic region" which denotes the functional groups in the molecule, and ν_{max} between 1350 cm^{-1} and 650 cm^{-1} (λ_{max} 7.4 nm to 15.3 nm) is called the "Fingerprint region" which is more important as two pure different organic compounds (enantiomers are excluded) will never exhibit same fingerprint in the spectra.

(d) The following Table and chart are used to intrerpret and locate the functional group of the molecule, the fingerprint region is used for comparison. The spectrum starts from higher frequency.

Table: ν_{max} cm^{-1} and (λ_{max} nm). approximate values

Sl. No.	CnHm	ArH	R–OH	–COOH	$–CO_2R$ –CO–O–	–C(=O)–	–CHO	CN	$C–NR_2$	$–NO_2$	Miscellaneous
											(a) Anhydride C=O 1870–1740 (5.44–5.75)
1.	–CH 3040–2870 (3.29 – 3.48)	–CH 3100–3000 (3.23 – 3.33)	Free OH 3640–3610 (2.77 – 2.75)	Bonded 3300–2500 (3.03–4.0)	C = O 1800–1635 (5.56–6.12)	C = O 1725–1705 (5.8 – 5.87)	–CH 2900–2700 (3.45 – 3.37)	C ≡ N 2260–2215 (4.43–4.52)	CH_3–N 2820–2760 (3.54 – 3.84)	1565–1335 (6.39–6.62)	(b) Acyl halide C=O 1815–1785 (5.51–5.60)
2.	– C = CH 3080–3020 (3.25–3.31)	1600–1450 (6.25–6.90)	Dimeric 3550–3450 (2.82–2.90)	Free 3560–3500 (2.81–2.86)	–C–O 1300–1100 (7.69–9.09)	Conjugated 1685–1665 (5.75–6.01)	Satd. C = O 1740–1720 (5.75–5.81)	$–CNH_2$ 3500–3100 (2.86–3.23)			(c) Weak signals for all C=O 3550–3200 (2.82–3.13) (d) Amide, C=O
3.	C ≡ CH ≈ 3300 (3.03)	Ar–CH_2– 1445–1430 (≈ 7)	Polymer 3400–3200 (2.94–3.13)	C=O 1700–1680 (5.88–5.95)		5-ring sat. (5.70–5.75)	Conjugd 1705–1680 (5.87–5.95)		–NH 1650–1550 (6.06–6.45)		1690–1630 –NH (5.92–6.14) 3500–3100 (2.86–3.23)
4.	C = C 1670–1640 (5.99–6.10)	Substd. Arom. compd. 800–700 (11.4–14.3)				6-ring sat. (5.80–5.87)					(e) S–H 2590–2550 (3.86–3.92) S=O 1070–1030
5.	C ≡ C 2260–2100 (4.43–4.76)					General (5.88 – 6.04)					(9.35–9.71) (f) Halogens C–Hal ≈ 750 ± 50 cm^{-1}.

(e) ν_{max} (or λ_{max}) may not be the strongest signal in the IR-picture which generally exhibits many sharp ν_{max} when pure. The ν_{max} means that this signal is detectable in its environment in the continuous trace of the IR-picture.

(f) The following chart shows the origin of the ν_{max} in a IR-picture by different functional groups (or bonds).

Chart (data in λ_{max}):

Sl. No.	Approx. λ_{max} nm	Groups (bonds) responding between λ_{max} 2.5 and 7.40 (Group frequency region), called "Characteristic region,"
1.	2.75–2.85	(a) –OH (free), (b) –COOH (monomeric)
2.	2.86–3.10	(a) R–C ≡ C–H, (b) –OH (H-bonded), (C) $–NH_2$/–NH– (free and bonded)
3.	3.11–3.34	(a) CH=CH, $CH=CH_2$, $C=CH_2$, (b) Aryl –C–H, (c) –OH (chelated), (D) Amino acid
4.	3.35–3.64	(a) CH_3, CH_2, CH, (b) –CO (dimer), (C) –COOH, (d) –OH (chelated), (e) Amino acids (d) Amino acids
5.	3.65–4.00	(a) OH (chelated).
6.	4.01–4.45	(a) R–C ≡ C–R, (b) –CN, (c) –NC, (d) –NCO
7.	4.46–5.00	(a) R–C ≡ C–H, (b) –NC, (c) Amino acids.
8.	5.01–5.40	(a) Anhydride (five-membered).
9.	5.41–5.74	(a) Ketone (four-membered), (b) Ester (Saturated, substituted), (c) γ-lactones, (d) Anhydride, (e) R–CO–Cl, (f) Acyclic and five ring anhydride.
10.	5.75–6.10 Very strong signal	(a) Ketone (five-membered), (b) Ketone (dialkyl and six-membered cycle), (c) α-Halogenated ketones, (d) R–CHO, (e) C = C–CHO, (f) C = C–C(=O)–C, (g) Ar–CHO, (h) R–COOH (i) C = C–CO_2H, (j) $ArCO_2H$, (k) –CO–NR_2 (R = H/alkyl group), (l) –CHO (chelated), (m) CH = CH_2 > C = CH, C = C, (n) C = C–C = C, (o) Aromatic ring, (p) Amines, (q) Amides.
11.	6.11–6.44	(a) β-Diketones, (b) –COO, (c) Amide.
12.	6.45–7.40	(a) Hydrocarbons, (b) Aldehyde, (c) Acid, etc.
13.	7.41–15.5	This is the "Fingerprint Region" (complex, difficult to designate all signals), characteristic absorption pattern of the entire molecule which is very useful in establishing the identify of the compound by comparing with known sample. Signals arise from various classes of organic compounds specially containing N, O, S, and halogens which are generally very prominent.

2.11 NUCLEAR MAGNETIC RESONANCE SPECTROSCOPY

To the organic chemist the nuclear magnetic resonance (n.m.r) spectroscopy is a very important tool. The detailed discussion of the topic is beyond the scope of this book. The beginners need not know the theoretical aspects of the origin of the spectrum. Only the ability for interpretations of the signals in the spectrum is required.

The nuclei of 12 C and 160 do not possess spin angular momentum, so they do not respond while 1H, 13C, 19F, 31P etc. (containing odd number of protons and or neutrons) furnish the signals. The proton magnetic resonance (p.m.r.) is most extensively used though the 13C n.m.r. is equally interesting.

The non-interfering solvents are CCl_4, CS_2, tetrachloroethene (all solvents are anhydrous) but, if needed, $CDCl_3$, and D_2O, etc. may be used taking care of their interference to some extent in the specific region in the spectrum.

The chemical environment of the magnetic nucleus is very important as the nucleus is exposed to different values of magnetic field thus absorbing at different frequencies. This is known as the chemical shift which helps the organic chemists to derive the structural formula of the given compound. The chemical shift is generally expressed as δ (delta; parts per million) and occasionally as τ (tau, $\tau = 10 - \delta$). The tetramethylsilane (TMS) is used as the internal standard, its twelve protons are equivalent and give a single line which is taken as the reference point in the spectrum and marked as $\delta = 0$. Factors like inductive effect and magnetic anisotropy effect influence the chemical shift. Chart I is helpful for the beginners, but is surely over-simplification of the complicated aspects of n.m.r. spectral analysis.

Chart I: Location of different types of protons or chemical shift region. Approximate positions of signals in the spectrum (600 Hz) shown by A, B, etc.

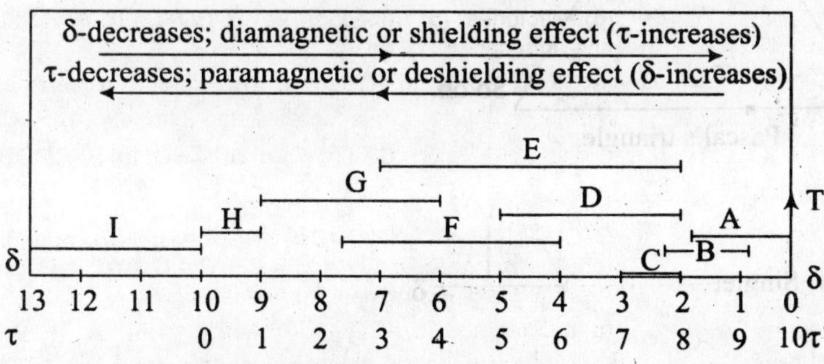

Class of compounds: I (Carboxylic acid, COOH), H (Aldehyde –**CH**= O), G (Aromatic and Heteroaromatic, Ar**H**), F (olefinic, C = **C**H), E(α-Disubstituted aliphatic), D(α-Mono-substituted aliphatic), C (Acetylenic), B(β-Substituted aliphatic), A (Aliphatic and/or Alicyclic) [A, B, D and E are saturated and C, F, G, H and I are unsaturated].

The Chart I shows that δ increases (i.e. paramagnetic shift) by (i) alkyl substitution, (ii) increased electromagnetic substitution, (iii) involved hydrogen bonding, etc. In the n.m.r. spectra, the following groups are readily recognized, $\mathbf{CH_3^-}$, CH_3-CH_2-, $CH_3-\mathbf{CH}<$, $Me_2\mathbf{CH}-$, $-\mathbf{CHO}$, $-\mathbf{OH}$, $>\mathbf{NH}$, $-O-\mathbf{CH_2}-O-$, $\mathbf{CH_3}-C\equiv C^-$, $-Ar-\mathbf{CH_3}$, $>N-\mathbf{CH_3}$.

The following points are to be noted:

(i) **Hydrogen atoms on the same carbon may not be equivalent in those molecules suffering restricted bond rotation.**

(ii) **The coupling constant, termed J (expressed in hertz) shows the separation (i.e. distance) between every pair of lines in the multiplet signals derived from the magnetic nuclei. The J value is independent of the applied magnetic field.**

(iii) **The molecules containing two types of protons (expressed as A–X system of protons) exhibit first-order splitting pattern, the multiplicity is given by (n + 1), where n is the number of equivalent nuclei. The relative intensity (height of each line in the multiplet (i.e. signal showing more than one line) is derived from the "Pascal's triangle" (shown below).**

1 ← single line (singlet, *s*,)

1—1 ← doublet (two lines of equal intensity, *d*)

1—2—1 ← triplet (three lines with intensity 1:2:1, *t*)

1—3—3—1 ← quarter (four lines with relative intensity 1:3:3:1, *q*)

1—4—6—4—1 ← pentet (five lines with relative intensity 1:4:6:4:1).

etc.

So on.

Pascal's triangle

(a) Singlet

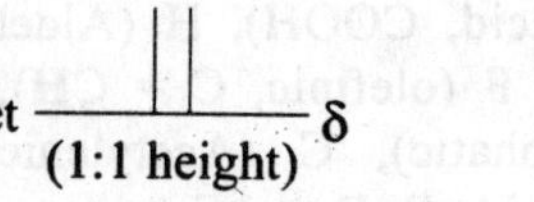

(b) Doublet

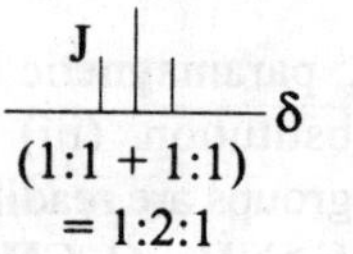

(c) Triplet

(d) Quartet 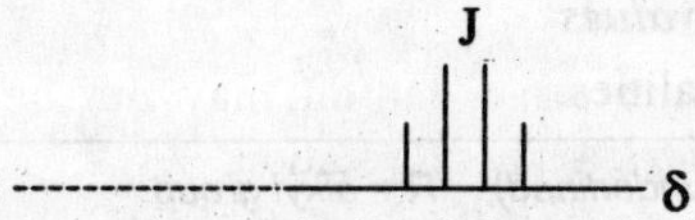

(1:1 + 2:2 + 1:1)
= 1:3:3:1

(e) Pentet 

(1:1+3 : 3+3:3+1:1)
= 1:4 : 6:4:1

(iv) The number of proton(s) giving signals at different δ values can be mechanically determined by integration. The relative hight of the integration curve at each signal is proportional to the number of proton(s).

e.g. (a) Ethanol, CH_3–CH_2–O–H (protons at different location is in the ratio 3:2:1). The n.m.r. picture looks like (vide "NMR Spectra Catalog.", Varian Associates; revised data, "Organic Chemistry" by I. L. Finar, Vol-2, 1975, p.32).

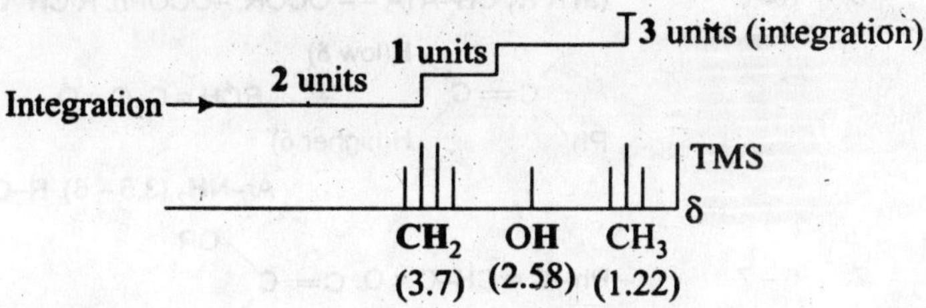

(b) Toluene, C_6H_5–CH_3
The n.m.r. picture looks like

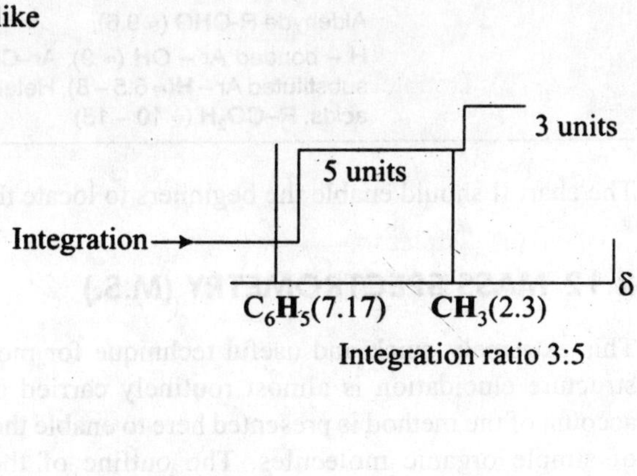

Integration ratio 3:5

Chart II: *Proton signals the δ-values*

H attached to different functionalities.

Sl. No.	*δ-values (Approx.)*	*H–atoms (underlined); R = alkyl group.*
1.	0–1	CH_3 – R, CH_3 – A (A = C–C = C, C –NHCOR, C–NR_2, –Cl) CH_2 in cyclopentane and cycohexane, etc.
2.	1–2	(a) CH_3 – A (A = –C = C, –C = C, –C ≡ CR, –CO_2R, – CN, –C–CO_2R, –C–$CONR_2$, –C–OR, –C–CN, –CNO_2, –C–X, X = Cl, Br, etc.). (b) R–CH_2–A (A = R′, –C = C, C–C = C, – CNR_2, C–I, –COR, C–CN, etc.). (c) RR_1 – CH – A (A = –R_2, –C – I, –C–NO_2, –C, Br, etc.) and H – C ≡ C
3.	2–3	(a) CH_3 – A (A = – CO.NR_2, –COOH, C–CHO, – SR, –I, –Br) (b) R–CH_2 – A (A = –CO_2R, –CN, –CO_2H, –SR, –CHO, etc) (c) RR_1 CH – A (A = –C = C, –CN, –CO_2H, – COR, –NH_2, etc). (d) Acetylene, H–C ≡ C–Ph, HC ≡ C – C = O, H – C ≡ C–C = C,
4.	3–4	(a) CH_3 – A (A = –OR, –OH, –OCOR, etc). (b) R–CH_2 – A (A = Br, – Cl, –I, –OR, –OH, etc). (c) RR_1CH – A (A = CO Ph, –Br, –Cl, –OR, – OH, etc).
5.	4 – 5	(a) CH_3 –A (A = –F, –NO_2, etc). (b) R– CH_2 –A (A = –CO Ph, –F, –NO_2, –O–COR, etc.). (c) RR_1CH – A (A = –NO_2). (d) CH_2 = CR_2, CH_2 = C–C = C, R – O·H, ArOH, R–NH_2 (δ 2.5),
6.	5 – 6	(a) R R_1 CH–A (A = – OCOR, –OCOPh. R·CH=C, Ph(–)C=C(–H(low δ))(–H(higher δ)) RCH = C–C = O Ar–NH_2 (3.5 – 6), R–$CONH_2$
7.	6 – 7	Ph–C = CH–C = O, C=C(–OR)(–H)
8.	7 – 13	*Miscellaneous:* Ph – CH = C = O (≈ 7.8), R–O–COH (≈ 8) Aldehyde R-CHO (≈ 9.6), H – bonded Ar – OH (≈ 9), Ar–CH = O (≈ 9.9), C = C–CH = O (≈ 9.8), substituted Ar – H(≈ 6.5 – 8), Heterocyclic, Het H(≈ 6.2 – 9.26), Carboxylic acids, R–CO_2H (≈ 10 – 13)

The chart II should enable the beginners to locate the protons of different molecules.

2.12 MASS SPECTROMETRY (M.S.)

This extremely quick and useful technique for molecular weight determination and structure elucidation is almost routinely carried out in many laboratories. A brief account of the method is presented here to enable the beginners to deduce the structure of simple organic molecules. The outline of the method which starts with the generation of molecular ion, $M^{+\cdot}$, is schematically shown below.

(A).

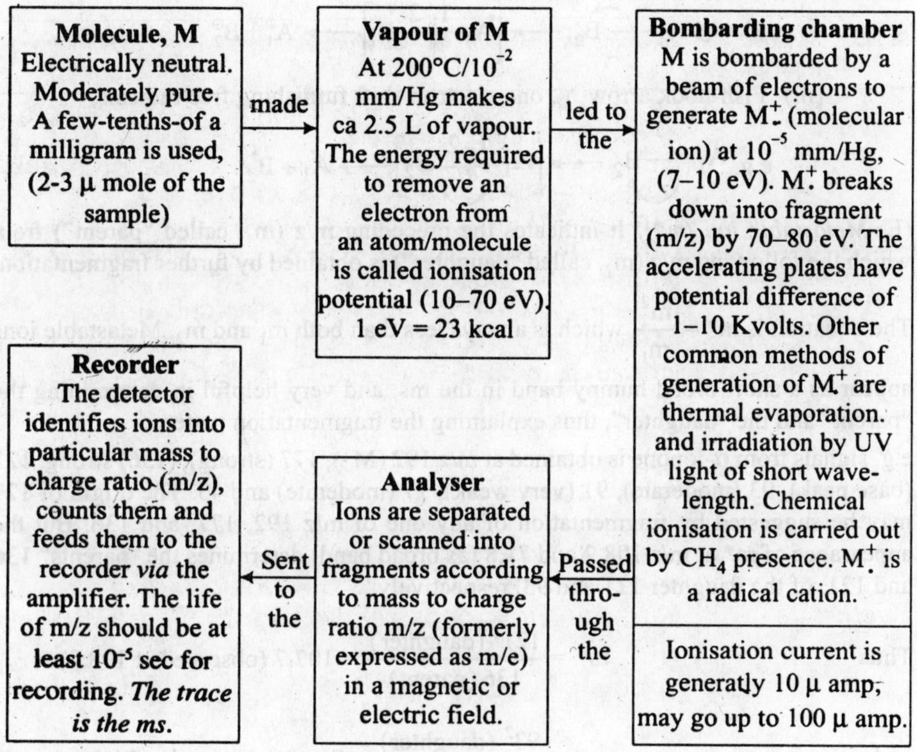

(B) *Equation:* $m/z = \dfrac{H^2r^2}{2V}$;

m = mass of the ion.

V = potential difference on the two accelerating plates.

r = radius of curvature of the path.

H = strength of the magnetic field.

(C) *Resolution:* (R). $R = \dfrac{M}{\Delta M}$;

M = mass of the particle.

ΔM = Difference between M and the next particle of mass (M + 1). Generally, R is ≈ 2,000, but sometimes it may go upto 10,000.

(D) *Symbols*

(i) Arrow sign, ↷, two-electron shift leads to generation of cation and anion.

(ii) $aC_x \overset{x\,\wr\wr\,y}{—} C_y$,. bond breaking between C_x and C_y, x < y (x comes before y). e.g.

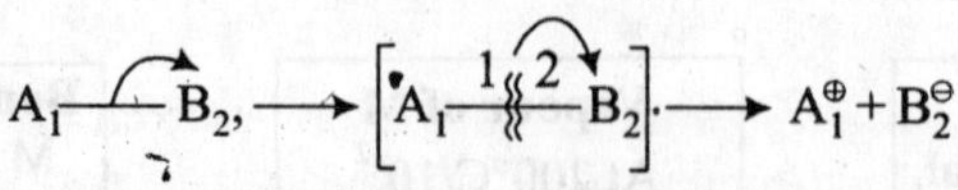

(iii) Fish-hook arrow ⇁, one electron shift furnishing free radicals,

e.g.

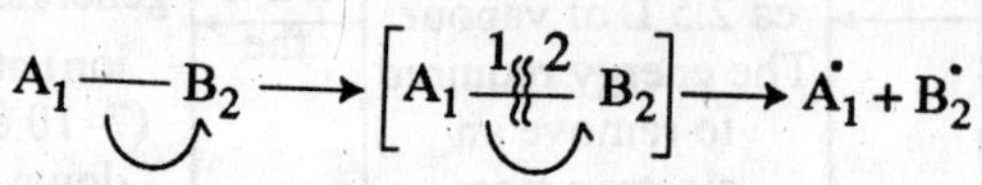

(E) *Metastable ion (m*)*: It indicates the preceding m/z (m_1, called "parent") from which the following m/z (m_2, called "daughter") is obtained by further fragmentation.

The equation is $m^* = \frac{m_2^2}{m_1}$ which is always less than both m_1 and m_2. Metastable ions appear as a short broad humpy band in the ms. and very helpful in determining the "parent" and the "daughter", thus explaining the fragmentation pattern.

e.g. signals from α-ionone is obtained at m/z 192 ($M^{\dot{+}}$), 177 (strong), (136) strong, 121 (base peak), 93 (moderate), 91 (very weak), 77 (moderate) and 43. The origin of 121 may be suggested by fragmentation of any one of m/z 192, 177, and 136. But the appearance of m* at m/z 108.2 and 71.8 (as broad band) determines the "parents" 136 and 121, of the daughter 121 and 93, respectively.

Thus $$m^* = \frac{121^2\ \text{(daughter)}}{136\ \text{(parent)}} = 107.7 \text{ (observed at 108.2).}$$

and $$m^* = \frac{93^2\ \text{(daughter)}}{121\ \text{(parent)}} = 71.47 \text{ (observed at 71.8).}$$

(F) *Base peak (b.p)*: The highest (the strongest) peak in the spectrum is called base peak and given an arbitrary value of 100. The base peak is derived from the most stable and more abundant cation. Other peaks at m/z is expressed as the percentage ratio of the base peak (called relative intensity, RI). The relative intensity of $M^{+\bullet}$ decreases roughly in the following order for different classes of organic compounds.

Aromatic/Heteroaromatic > Conjugated olefin > Alicyclic > Sulphide > Unbranched hydrocarbon > Mercaptan > Ketone > Amine > Esters > Ethers > Carboxylic acid > Branched hydrocarbon > Alcohol.

Thus a strong $M^{\dot{+}}$ Signal is not expected of an alcohol or amine, likewise a very weak $M^{\dot{+}}$ signal is not expected from an unsubsituted aromatic compound.

(G) *The following points are noteworthy:*

(a) In the ms the true mass is obtained. Small eV gives fewer m/z signals and strong $M^{\dot{+}}$ which is the highest mass (isotopic signals of very low intensity are noted sometimes). The $M^{\dot{+}}$ contains odd number of electrons and called radical cation. It undergoes further fragmentation, the extent depends on the nature of the $M^{\dot{+}}$ In $C_nH_mN_xO_yS_z$, where x = even number (viz. 0, 2, 4 ... etc.) the $M^{\dot{+}}$ is even number but when x = odd number (viz. 1, 3, 5 ... etc.) the $M^{\dot{+}}$

is odd, n, m, y and z are immaterial. The nature of the m/z (odd or even) depends on the portion eliminated as radical or neutral molecule from the initially formed $M^{+\cdot}$.

(b) The same signal at m/z ratio may be obtained from differnt fragments but their RI may differ. Isotope labelling would shift the position of the m/z signal of the fragment under consideration, e.g. $C_3H_7^+$ (m/z 43) and $C_2H_3O^+$ (m/z 43) give signals at the same m/z ratio, but the former ion is shifted to m/z 44 (for $C_3H_6D^+$) when D-Substituted compound is taken. Similary C_2H_5–CH = O gives a signal at m/z 29 which may be due to either $C_2H_5^+$ or CHO^+ (or both). But when $C_2H_5CD = O$ (M) is taken, then only m/z 30 signal is obtained (due to $CD = O^{\oplus}$) proving that the m/z 29 was obtained solely from $CHO]^+$ (and not from $C_2H_5^+$) in C_2H_5CHO fragmentation.

(c) The smallest m/z (particularly stabilised by resonance) tends to be the base peak (b.p.).

e.g. $EtCH(NH_2)CH_3]^{+\cdot} \longrightarrow H_2\overset{\oplus}{N}{=}CHMe + H_2\overset{\oplus}{N}CHEt + H_2\overset{\oplus}{N}{=}CMe(Et)$

[$M^{+\cdot}$, m/z 73] — Strong RI bp. m/z 44 — Weak RI m/z 58 — V. Weak RI m/z 72

(d) The stability of the liberated particle (radical or molecule) is also a contributing factor toward the generation of stable fragment ions. The six-centred rearrangement of the $M^{+\cdot}$ in the fragmentation (McLafferty rearrangement) is to be carefully considered as a possibility. The general form of the McLafferty rearrangement is shown below. This McLafferty rearrangement is very sensitive of the stereochemistry of the molecule.

(i)

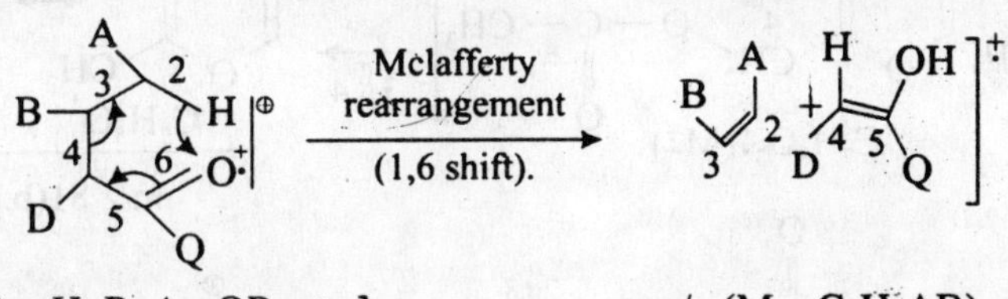

[Q = H, R, Ar, OR, etc.] — m/z ($M - C_2H_2AB$)

(ii)

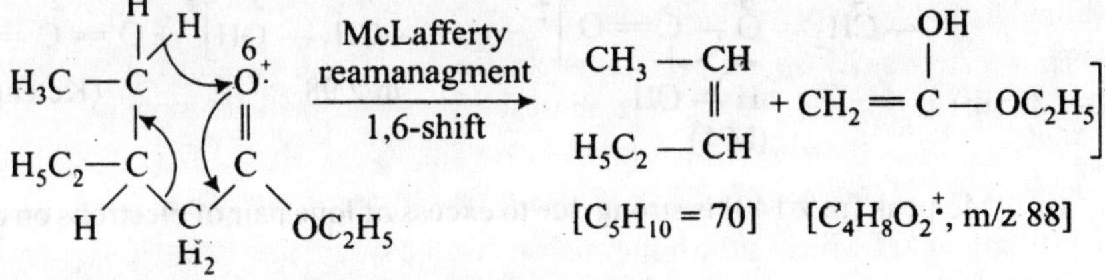

[C_5H_{10} = 70] [$C_4H_8O_2^{+\cdot}$, m/z 88]

[$C_9H_{18}O_2]^{+\cdot}$ m/z 158 ($M^{+\cdot}$)]

(e) The speculated structure derived from the ms analysis should be further corroborated by elemental analysis, uv, IR, nmr spectra (as many as possible).

Representation of picture: (over-simplified), m/z (RI).

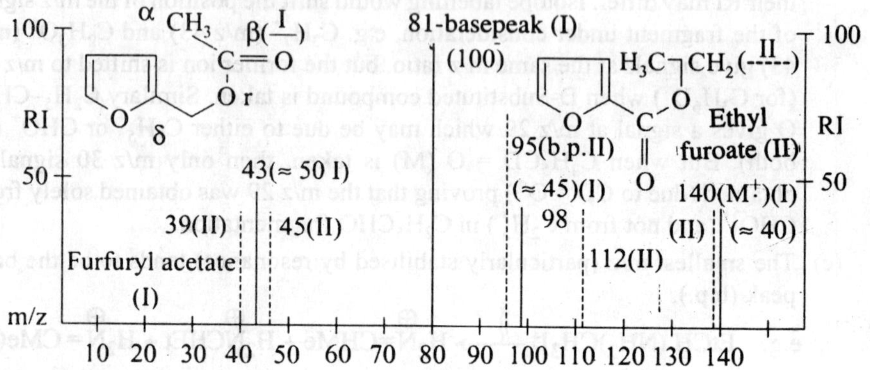

(*The RI's are approximate*): Broken line for ethyl furoate. Minor signals at $(M + 1)^{+}$ and $(M + Z)^{+}$ are not shown.

(b) *Presentation:* m/z (RI): M^{+} (140 (≈ 40), m/z 98 (≈ 45), m/z 81(100), m/z 43(≈ 50) for I. M^{+} 140 (≈ 40), m/z 112 (≈ 30), m/z 95 (100), m/z 45 (≈ 20), m/z 39 (≈ 45) for II.

(c) *Assignments:*(I)

(i)

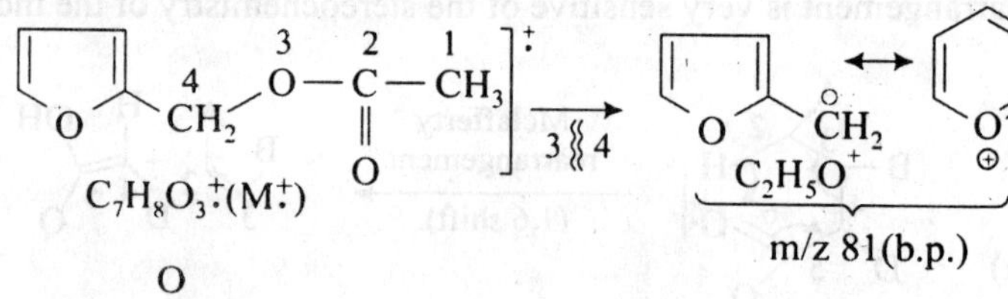

(ii)

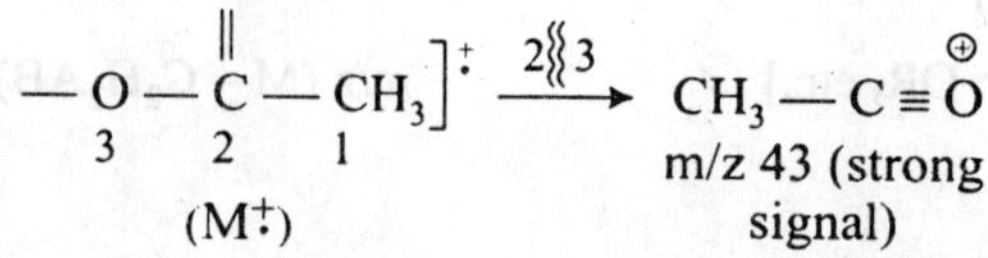

(iii)

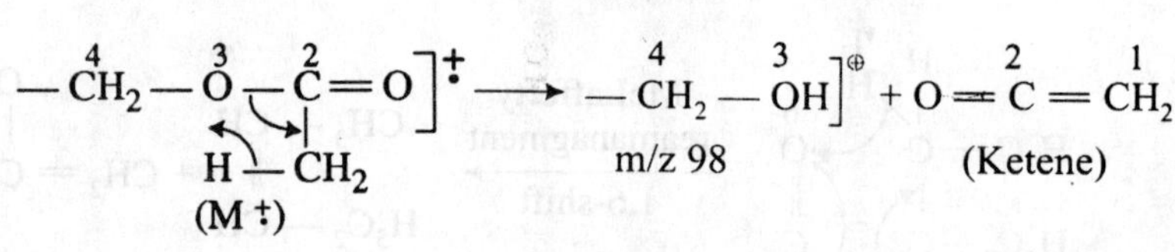

(iv) M^{+} peak (m/z 140) is strong due to excess of lone pair of electrons on oxygen.

Chart:

Sl.No.	Class of compounds	Major fragmentation pattern, m/z (RI).
1.	*Alkane* $\left[-\overset{\vert}{\underset{\vert}{C}}-\overset{\vert}{\underset{\vert}{C}}-\right]^{+\cdot}$ ($M^{+\cdot}$) $\longrightarrow -\overset{\vert}{\underset{\vert}{C}}^{\oplus} + \cdot\overset{\vert}{\underset{\vert}{C}}-$ (m/z) $M^{+\cdot}$ (v.w. with increasing mol. wt.); identification becomes difficult with increasing mol. wt. and branching (tends to give higher degree of $C^{\oplus}$). Usually the b.p. is at C_3/C_4.	 $]^{+\cdot} \longrightarrow$ m/z(99, w), 85 (w), 71 (w) (57 s), 43 (b.p.). e.g. (a) (octane, C_8H_{18} M±(114)w [*Note*: $M^{+\cdot} \xrightarrow[(-15)]{-CH_3^{\cdot}}$ m/z $\xrightarrow[(-14)]{-CH_2^{\cdot}} \rightarrow$ b.p.] 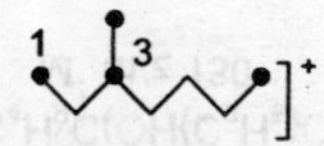 $]^{+} \longrightarrow$ m/z(99, w), 85 (w), 57($C_4^{\oplus}$, b.p.) (b) (3-Methylheptane C_8H_{18}, M^{+} 144)(w.) [*Note*: branching at C_3 generates $2°C^{\oplus}$] (c)  $\longrightarrow$ m/z (99, w), 85 (w), 71 (b.p.) (3,3-Dimethylhexane C_8H_{18}, M^{+} 114 (v.w.) [*Note*: Fragmentation at C_3 generates very static $3°C^{\oplus}$ making b.p. $C_5H_{11}^{\oplus}$]
2.	*Alkenes:* Isomeric olefins give similar spectra. $-C_3{=}C_2{-}C_1^{\oplus} \longleftrightarrow {-}C_3^{\oplus}{-}C_2{=}C_1{-}, M^{+\cdot}$ unstable, generates rapidly stable allylic $C^{\oplus}$ by fragmentation.	The RI of m/z decreases with increasing molecular weight of the M^{+}. e.g. $[R-C-C{=}C-R_1]^{+\cdot} \xrightarrow{-R^{\cdot}} \overset{\oplus}{C}-C{=}C-R_1 \longleftrightarrow C{=}C-\overset{\oplus}{C}-R_1$ *major* (M — R) $M^{+\cdot}$ $R > R_1$; So R is lost to a greater extent $\xrightarrow{-R_1^{\cdot}} R-C{=}C-\overset{\oplus}{C} \longleftrightarrow R'-\overset{\oplus}{C}-C{=}C$ *minor* (m – R_1)

3.	Substituted benzene and heteroaromatic compounds.	(a) $Ph-CH_2-CH_2-R_1]^{\dot{+}} \xrightarrow{\wr\wr} Ph-CH_2^{\oplus} \longrightarrow C_7H_7^{\oplus}$ (cyclic) m/z 91 b.p. (stable, resonance)
	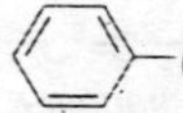 $-(CH_2)_n-R]^{\dot{+}}$ $M^{\dot{+}}$ is very weak and unstable.	(b) 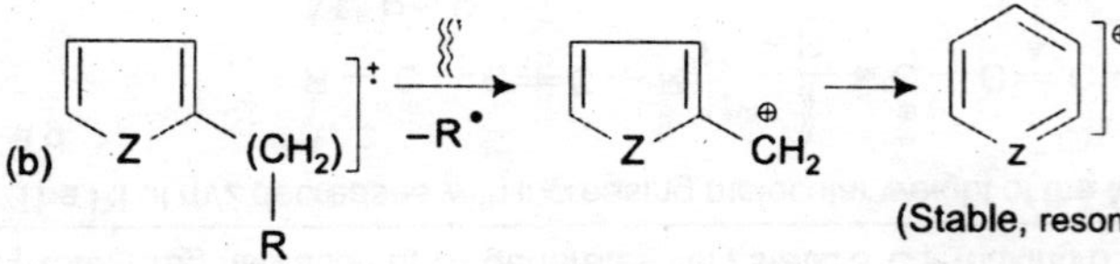 (Stable, resonance)
4.	*Substituted alkanes:*	*Stability of m/z:* (i) smallest m/z > large m/z.
	$R-C(R_1)(R_2)-Z]^{\dot{+}} \rightarrow -\overset{\mid}{\underset{\mid}{C}}^{\oplus} + Z^{\cdot}$ $M^{\dot{+}}$ (W.)	(ii) $C^{\oplus}$ on α-carbon of N > S > O > Hal.
	[R > R_1 > R_2; Z = Hal, –S–, O– N<]	(iii) $3°C^{\ominus} > 2°C^{\oplus} > 1°C^{\oplus}$
	Ethers undergo similar fragmentation at α position	(iv) $N > Ph-\overset{\oplus}{C}H^2 > O$
		(a) $C_4H_9C(OH(C_3H_5)CH_3)]^{\dot{+}} \xrightarrow[-C_4H_9]{\wr\wr} \overset{+}{C}(OH)(C_2H_5)(CH_3)$ $M^{\dot{+}}$ m/z 130 b.p. m/z 73.
	$R-CH_2-O-R_1]^{\dot{+}} \rightarrow R_1-\ddot{O}-CH_2^{\oplus}$	(b) $R-CH-CH_2-\overset{\oplus}{C}HOH$ (with H) $\longrightarrow$ Alkene + $\overset{\ominus}{C}H_2(OH)$ m/z 31
5.	Alcohols, amines, thiols R–C–C–OH, R–C–C–NH_2 R–C–C–SH	gives m/z (19, $H_3O^{\oplus}$), m/z (18, $NH_4^{\ominus}$), m/z (35, $H_3S^{\oplus}$), respectively.

(II)

(i) 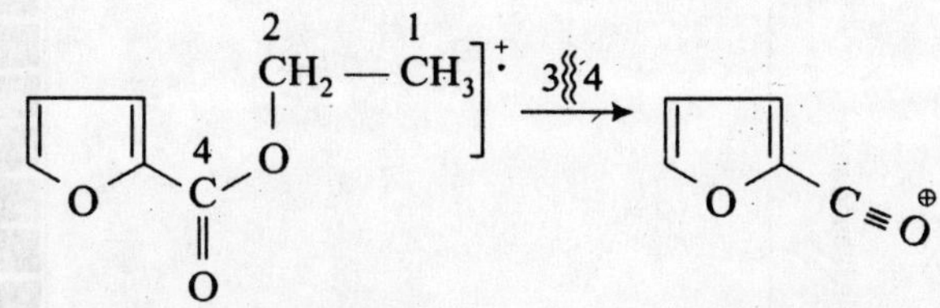

M$^{+\cdot}$, (m/z 140, strong) m/z 95, (bp)

(ii) M$^{+\cdot}$ $\xrightarrow{3 \wr\wr 4}$ $\overset{\oplus}{O}$ — CH_2 — CH_3

(m/z 45, weak)

(iii) 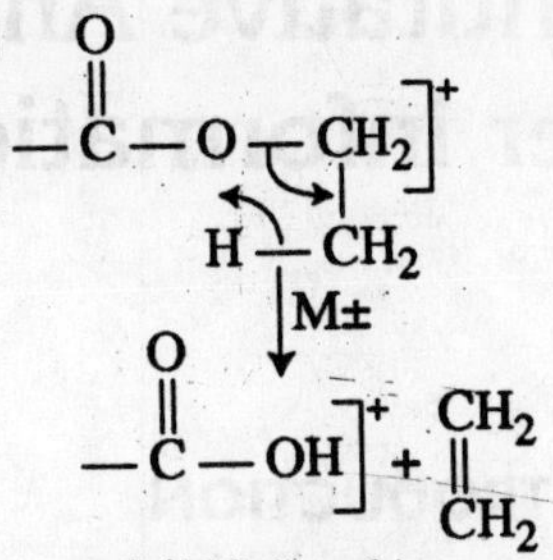

m/z 112 (weak)

(iv) M$^{+\cdot}$ ⟶ $C_3H_3^{\oplus}$

m/z 39 (w)

(d) Note that isomeric compounds can be distinguished by ms from their different fragmentation pattern.

(I) *Typical fragmentation pattern of different classes of organic compoud:* [RI of fragment shown as b.p. (base peak), v.w (very weak). w (weak), m (moderate ≈ 50%), s (strong), v.s. (very strong ≈ bp) in the given chart.

Quantitative Analysis and Other Information

3

3.1 INTRODUCTION

The analysis of the organic compounds is also an essential part of the practical organic chemistry. The synthetic organic chemists know quite well that without the elemental analysis (C, H and N, mainly) the proposed identity of the compound (prepared first time) remains unacceptable. Similary in the natural product chemistry, the compounds isolated from plant or animal source need the elemental analysis and determination of molecule weight to assign the chemical formula from which the structure elucidation of the natural products starts. So the responsibility of the analysts is quite high. Their manipulative skill should be excellent and they must love this sort of monotonous and demanding work for years together. They have the extra responsibility to train up the beginners to fill up the vacant positions when they retire.

3.2 THE ANALYSIS OF CARBON AND HYDROGEN IN PURE ORGANIC COMPOUNDS

The detailed description of the apparatus set-up and procedure is not necessary here because it is highly desirable to learn this part directly from the supervisors (analyst). This will save lots of time and labour as well as embarrassment. Very rarely the investigators themselves analyse their products (synthetic/natural). On the contrary, they submit the products to the analyst to get the unbiased analysis. However, the chemistry of the analysis should be known to them. This is given below. The analysis should involve pure reagents and solvents in all experiments.

Oxygen from the cylinder is bubbled through alkali (to free it from CO_2) followed by conc. H_2SO_4 (to free in from moisture) very slowly (1 bubble per sec.) and passed through the combustion tube containing the weighed sample in a porcelain boat. The combustion tube heated slowly at the beginning, then the temperature is increased to red heat when the sample is burnt quantitatively to furnish CO_2 and H_2O (vapour).

Special precaution is taken to arrest the unwanted acidic gases like SO_2, oxides of N etc. in the combustion tube by choice of suitable packing materials so that only CO_2 and H_2O vapour can emerge. The issuing gas mixture is passed successively through (i) a tared U-tube containing conc. H_2SO_4 (H_2O absorbed) soaked over pumice stone-chips, (ii) tared U-tube containing NaOH on asbestos (CO_2 absorbed) and finally (iii) a bubbler containing conc. H_2SO_4 as a guard tube (to protect the absortion of moisture from the atmosphere by the highly hygroscopic contents of the second U-tube). Generally the experiment is completed within an hour then the U-tubes are dismantled, weighed and the amounts of H_2O and CO_2 formed from the combustion of the weighed sample are determined by differences in the initial and final weights of the U-tubes. The following formulas are used for the calculations of %C and %H (A and B) in the sample: [W_S = weight of the sample in gm, W_{CO_2} = wt. of CO_2 and W_{H_2O} = wt. of H_2O in gm, formed after combustion of the sample].

$$\text{[A] \%C} = \frac{W_{CO_2} \times 12 \times 100}{W_S \times 44} \; or \; \frac{W_{CO_2}}{W_S} \times 27.27 \text{ (the formulation is obvious)}$$

$$\text{[B] \%H} = \frac{W_{H_2O} \times 2 \times 100}{W_S \times 18} \; or \; \frac{W_{H_2O}}{W_S} \times 11.11 \text{ (–do–).}$$

I. The analyses are qualified as "Macro", "Semimicro", and "Micro", depending on the weight of the sample taken for each analysis.

(i) Macro, when the wt. of sample is 0.xyza gm ($x \neq 0$), generally 0.1-0.2 gm.

(ii) Semimicro, when the W_S is 0.0 yza gm ($y \neq 0$), generally 0.03-0.04 gm.

(iii) Micro, when the W_S is 0.00 za gm ($z \neq 0$), generally 0.003-0.005 gm. It is desirable to use a chemical balance sensitive upto fifth place of decimal while conducting the micro-analysis.

II. Examples:

(i) Macro: **Weight of the sample = 0.2701 gm, obtained CO_2 =0.3926 gm and H_2O = 0.1621 gm. Putting these values in the previous equations,**

$$\%C = \frac{W_{CO_2}}{W_S} \times 27.27 = \frac{0.3962 \text{ gm}}{0.2701 \text{ gm}} \times 27.27 = 40.00.$$

$$\%H = \frac{W_{H_2O}}{W_S} \times 11.11 = \frac{0.1621 \text{ gm}}{0.2701 \text{ gm}} \times 11.11 = 6.66.$$

In the absence of other information it may be assumed that the sample contains oxygen since the total of %C and %H does not account for 100%. So the sample contains (100 – 40.00 – 6.66) % 0 = 53.34 % oxygen in the molecule.

(ii) Semimicro: **W_S = 0.0432 gm. W_{CO_2} = 0.1230 gm, and W_{H_2O} = 0.0279 gm.**

Therefore, the $\%C = \frac{W_{CO_2}}{W_S} \times 27.27 = \frac{0.1230 \text{ gm}}{0.0432 \text{ gm}} \times 27.27 = 77.64,$

and the $\%H = \frac{W_{H_2O}}{W_S} \times 11.11 = \frac{0.0279 \text{ gm}}{0.0432 \text{ gm}} \times 11.11 = 7.17.$

(77.64 + 7.17)% = 84.81%, so the remaining portion is oxygen,

Therefore, %O = (100 – 84.81) = 15.19.

(iii) Micro: $W_S = 0.00466$ gm, $W_{CO_2} = 0.01560$ gm, $W_{H_2O} = 0.0034$ gm.

Therefore, $\%C = \frac{W_{CO_2}}{W_S} \times 27.27 = \frac{0.01560 \text{ gm}}{0.00466 \text{ gm}} \times 27.27 = 91.29$

and $\%H = \frac{W_{H_2O}}{W_S} \times 11.11 = \frac{0.0034 \text{ gm}}{0.00466 \text{ gm}} \times 11.11 = 8.10.$

These weights may be converted into milligram to make the calculations simplified, Thus,

$\%C = \frac{15.60 \text{ mg}}{4.66 \text{ mg}} \times 27.27 = 91.29$ and $\frac{3.4 \text{ mg}}{4.66 \text{ mg}} \times 11.11 = 8.10.)$

Since the total of %C and %H is (91.29 + 8.10) = 99.39, the remaining 0.61% is ignored. [The units of weights cancel during division]

3.3 DETERMINATION OF %N IN THE PURE ORGANIC SAMPLES

[A] Depending on the constitution and property of the nitrogenous sample, two methods have been developed for the estimation of nitrogen. The first method is named *Duma's method* (applicable to wide range of nitrogen containing compounds) involving the generation of elemental N_2 gas which is volumetrically measured. Thus the pure and dry sample is accurately weighed (generally semimicro, but occasionally micro), mixed intimately with CuO, taken in a tube and heated strongly when a mixture of nitrogen and oxides of nitrogen is formed together with CO_2 and H_2O vapour. This gas mixture is driven by pure CO_2 and passed through strongly heated Cu powder packed at the exit end of the combustion tube and the issuing gas mixture is passed through a strong solution of KOH which absorbs all the acidic gases and N_2 is collected in the nitrometer as microbubbles over the aqueous alkali. After the completion of the experiment the volume of the collected N_2 is noted, converted the volume at STP, aqueous tension at room temperature (t°C) is noted, and the %N in the sample is calculated in the following way:

(i) Let V_1 ml of N_2 is collected over alkali solution at t°C and P_1 mm/Hg pressure. If the aqueous tension at t°C is found to be f, then the actual pressure will be $(P_1 - f)$ mm/Hg and T_1 is (273 + t)°C.

Therefore, the volume of N_2 at STP (V_2 ml) is calculated as:

$$\frac{(P_1 - f) \times V_1}{T_1} = \frac{P_2 V_2}{T_2} \text{ or } V_2 \text{ (vol. at STP)} = \frac{(P_1 - f) \times V_1 \times 273}{(273 + t) \times 760} \text{ ml}$$

(ii) Since 22.4 L of N_2 weighs 28 gm at STP, therefore, V_2 ml of N_2 at STP will weigh $\frac{28 \times V_2}{22400}$ gm. Putting this value of the weight of obtained N_2 in the previous equation (step i) the new equation becomes,

$$\text{weight of } N_2 (W_N) = \frac{28 \times (P_1 - f) \times V_1 \times 273}{22400 \times (273 + t) \times 760} \text{ gm.}$$

(iii) If the weight of the sample taken is W_S gm. then %N will be $\frac{W_N \cancel{gm}}{W_S \cancel{gm}} \times 100.$

Putting the values of W_N (obtained in step ii) the final equation stands as [A].

$$\%N = \frac{28 \times (P_1 - f) \times V_1 \times 273 \times 100}{W_S \times 22400 \times (273 + t) \times 760} \text{ or } \frac{V_1 (P_1 - f) \times 28 \times 273 \times 100}{W_S \times (273 + t) \times 760 \times 22400}.$$

$$\text{So, } \%N = \frac{V_1 (P_1 - f)}{W_S (273 + t)} \times 0.0449 \qquad \text{...[A]}$$

Example: Given that:
$W_S = 0.4308$ gm, $t = 20°C$, $P_1 = 740$ mm (aq. tension at 20°C = 17.5 mm)
$V_1 = 45.70$ ml. What is the %N in the sample?

Ans: According to the working equation [A], the %N is calculated as,

$$\%N = \frac{V_1 (P_1 - f)}{W_S (273 + t)} \times 0.0449 \qquad \text{...[A]}$$

Putting the given data in the eqn. [A],

$$\%N = \frac{45.70 \times (740 - 17.5)}{0.4308 \times (273 + 20)} \times 0.0449 \text{ or } \frac{45.70 \times 722.5 \times 0.0449 \cancel{gm}}{0.4308 \times 293 \cancel{gm}}$$

= 11.74. So the sample contains 11.74%N. (*Ans*).

[B] Estimation of Nitrogen by Kjeldahl Method

(i) [Pronounced as Kyeldahl, Keldahl or Jeldahl.].
(ii) Macro (0.2–0.3 g), Semimicro (20–40 mg).

[C] Outline of the Determination (in Flowsheet)

Accurately weighed Nitrogenous compound (supplied sample) (not suitable for hetrocyclic compounds) —[AR. $H_2SO_4 + \Delta$, digested ca 45 min triple catalyst. ($CuSO_4$: K_2SO_4: Se in 6:2:1 by wt.)]→ $(NH_4)_2SO_4$ (Product) Quantitative —[Excess of alkali Δ]→ Evolved NH_3 (absorbed in known excess of standard H_2SO_4). Quantitative

[D] Reagents

(All reagents are of analytical grade i.e. A.R).

(i) Anhydrous Na_2CO_3 (standard solution):	A solution of nearly 0.02 (N) strength (standard solution) is made from accurately weighed Na_2CO_3 (0.265 g /250 ml = 0.02 N).
(ii) Conc. H_2SO_4 (3 ml for the digestion):	A solution of nearly 0.02 (N) strength 3 drops/250 ml = 0.02 N (*to be standardised* by the standard Na_2CO_3 solution)
(iii) NaOH (solid):	A strong solution is made from NaOH in H_2O (1:1, w/w) (This is about 40% NaOH solution; 10 ml to 12 ml used).
(iv) Triple catalyst:	$CuSO_4$ (6 parts by wt;, helps decomposition of the compound), + K_2SO_4 (2 parts by wt.; helps raising the b.p. of conc. H_2SO_4), + Se (1 part by wt.; helps dehydrogenation and oxidation), about 0.5–1 g of the catalyst is used in each experiment.
(v) Freshly boiled D.D. H_2O:	Used throughout the determination in reagent preparations, as solvent and washings.
(vi) Wesslow's indicator:	Nearly 250 mg of methyl red in 100 ml of pure ethanol is mixed with about 200 mg methylene blue in 100 ml of pure ethanol to furnish the indictor. Two drops of this indicator is used in each titration, the end point is very sharp (changing purple to green).

[E] Procedure

(For Semimicro determination): First experiment should be done under the supervision of the teacher.

(i) In a clean, dry and flat bottomed glass ampule, able to stand erect on the weighing pan, is taken accurately weighed (20 – 30 mg) sample and covered with nearly 0.5 g of the triple catalyst. This ampule is carefully introduced in the Kjeldahl flask (bulb nearly 50 ml), by slanting the neck (the contents should not stick to the neck and the ampule should not hit the bottom of the flask hard enough to break it) and treated with conc. H_2SO_4 (3 ml; a long neck dropper is used to deliver the acid at the bottom of the flask, the ampule should be under the acid level).

(ii) The reaction mixture is slowly heated over a small flame for 10 mins then strongly for 45 mins. An inverted clean funnel connected to the water aspirator is mounted on the open neck of the Kjeldahl's flask (now called digestion flask) to prevent environment pollution by the evolved SO_2 gas. When the decomposition is complete, the solution becomes transparent and colourless ($CuSO_4$ becomes anhydrous by conc. H_2SO_4) which is cooled and treated with H_2O (5 ml), the solution turns pale green ($CuSO_4 \cdot 5H_2O$ regenerated) and remains clear (care : H_2O on H_2SO_4).

(iii) The steam is generated (to expel dissolved NH_3, if any). The distillation set is connected with the digestion flask, the conc. NaOH solution (Reagent, iii, 20 ml) is taken in the cup on the upright distillation flask, the outlet of the condenser is connected to a piece of glass tube (clean) by a rubber tubing (if required) to reach the surface of the dilute standard (0.02 N order) H_2SO_4 (Reagent ii) solution (25 ml, pipette) to which is added 2 drops of Wesslow's indicator (to ensure indication of acidity of the acid solution throughout the distillation process) taken in an Erlenmeyer flask (clean, 250 ml). Two or three blocks of wood may be used as the platform of the Erlenmeyer flask, one after another may be removed, as and when required, to keep the depth of the condenser outlet at the optimum level (2 – 3 mm) to prevent the suction of a large volume of standard acid to spoil the experiment. Water is passed through the condenser (bottom inlet, up outlet), the steam outlet (moderate flow, medium flame) is connected to the distillation set-up via a rubber tubing holding a pinch cock (to control the flow of steam). About 8 ml of NaOH solution is run in the distillation flask (the remaining 1–2 ml acts as the seal).

(iv) The digestion flask is heated on a small flame protected by a flame-guard (so that the flame is not blown out by strong wind). The evolving NH_3 is chased by the steam into the H_2SO_4 solution contained in the Erlenmeyer flask. The distillate (50–60 ml) is collected into the Erlenmeyer flask (ca 45 mins) which is gradually lowered by removing the wooden platform as the necessity demands to protect from suction backward. The indicator must show the acidity of the resultant solution.

(v) The condenser is disconnected, inside of the condenser is washed down with little water into the Erlenmeyer flask, outside of the glass tube or condenser dipped into the acid is washed down with little boiled out water. The cold residual H_2SO_4 is titrated against the standard Na_2CO_3 (0.02 N order), if required 1 drop of Wesslow's indicator may be used again.

(vi) The steam generator may be cut-off from the source of heat, and the short flame heating the reaction mixture in the digestion flask is removed. If the flame cannot be made short enough, it could be placed a little away from the digestion flask. A sufficiently long glass tube should be used on the top of the steam generator to release excess of pressure developed inside it.

[F] Calculation

(i) $$\%N = \frac{(V_1 \times x \cdot \cancel{N} - V_2 \times y \cdot \cancel{N})1.4}{W_g \times \cancel{N} \times \cancel{g}}$$

[Cancellation of units in shown].

Working equation.

$$\%N = \frac{N(V_1 \times x - V_2 \times y) \times 14g \times 100}{Wg\ gm \times 1000\ N}$$

(actual equation)

V_1 ml = Volume of H_2SO_4 taken for absorption of NH_3

x = Normality of the H_2SO_4 solution (almost 0.02 N)

V_2 ml = Volume of the base used to neutralise the residual H_2SO_4 in the Erlenmeyer flask.

y = Normality of the base used to neutralise the absorbing H_2SO_4

W_g = Weight of the sample taken in gram.

Alternately,

$$\%N = \frac{(V_1 - V_2)N_b \times 1.4}{W_g}$$

V_1 ml = Volume of the base (standard solution) required to neutralise Z ml of dil. H_2SO_4 for standardisation (generally, Z = 25 ml).

V_2 ml = Volume of the base required to naturalise the residual dil. H_2SO_4 after NH_3 absorption.

N_b = Normality of the base (Na_2CO_3), (nearly 0.02 N).

W_g = Weight of the sample in gram.

NOTES:

(i) The glass ampule should be small enough to hold the sample and catalyst yet may be submerged under the conc. H_2SO_4 (3 ml). If required the digestion flask may be swirled to distribute the contents in the conc. H_2SO_4 for unhindered digestion.

(ii) The weighings of sample and Na_2CO_3 may be done simultaneously so that while the digestion is in progress, the dil. H_2SO_4 may be prepared and standardised by the standard base. Wesslow's indicator gives sharp end point.

(iii) Conc. H_2SO_4 (3 ml) should be measured from the burette into a small Erlenmeyer flask (10 ml) and delivered into the bulb of the digestion flask by a long-neck dropper.

(iv) A drop of pure water may be added to each standard joint of the distillation assembly to prevent leakage during distillation. The whole set-up must be checked for leak-proof.

(v) During the distillation by steam constant attention must be paid to ensure proper rate of chasing, to maintain undisturbed boiling and take necessary measures to prevent sucking back of the acid solution from the Erlenmeyer flask.

Example:

(i) $W_g = 0.2159$ gm. The ammonia produced neutralised 27.4 ml of 0.1 (N) H_2SO_4 soln. Calculate the %N in the sample.

Ans: 27.4 ml of 0.1 N $H_2SO_4 \equiv 2.74$ (N) $H_2SO_4 \equiv 2.74$ (N) NH_3 soln. Therefore, from the working equation [A] the %N becomes,

$$A = \frac{[V_1 \cdot x - V_2 \cdot y)}{W_g} \times 1.4, \text{ so } \%N = \frac{[2.74] \times 1.4}{0.2159} = 17.76\ (Ans).$$

(ii) (Different form of the (i) above)–

$W_g = 0.2159$ gm. The NH_3 produced is passed into 50 ml of 0.1 (N) H_2SO_4 soln. The residueal H_2SO_4 required 45.2 ml of 0.05 (N) NaOH soln. What is the percentage of N in the sample?

Ans: Given $W_g = 0.2159$ gm, H_2SO_4 used = 50 ml × 0.1 (N), and NaOH required to neutralise excess of H_2SO_4 = 45.2 ml × 0.05 (N). Therefore, from the working formula, [A],

$$\%N = [A] = \frac{[V_1.x(N) - V_2.y(N)]}{W_g[N]} \times 1.4, \text{ so } \frac{[50 \times 0.1 - 45.2 \times 0.05]}{0.2159} \times 1.4$$

$$\text{or } \frac{[5 - 2.26] \times 1.4}{0.2159} = \frac{[2.74] \times 1.4}{0.2159} = 17.76\ (Ans).$$

(iii) Given $W_g = 0.1074$ gm, H_2SO_4 soln. taken = 50 ml of 0.12 (N) strength, the residual H_2SO_4 required NaOH soln. = 109.6 ml of 0.041 (N) strength for neutralisation, what is the percentage of N in the sample?

Ans: Utilising the working equation for the determination of %N in the sample it is obtained that,

$$A = \frac{[V_1.x(N) - V_b.y(N) \times 1.4 \text{ gm}}{W_S \text{ gm} \times (N)} = \frac{(N)(V_b.x - V_b.y) \times 1.4 \text{ gm}}{(N) \times W_S \text{ gm}} = \%\ N.$$

Putting the given values in the equation [A], it is obtained,

$$\%N = \frac{(50 \times 0.12 - 109.6 \times 0.041) \times 1.4}{0.1074} = \frac{(6.0 - 4.49) \times 1.4}{0.1074} = 19.64\ (Ans).$$

(iv) Given $W_S = 0.5371$ gm, H_2SO_4 taken = 50 ml of 0.1(N); NaOH required to neutralise the residual H_2SO_4 = 45.2 ml of 0.05 (N). What is the %N?

Ans: From the working equation [A], it is obtained

$$[A] = \frac{[V_a.x - V_b.y]}{W_S} \times 1.4, \text{ so } \%N = \frac{[50 \times 0.1 - 45.2 \times 0.05] \times 1.4}{0.5371}$$

$$= \frac{(5 - 2.26) \times 1.4}{0.5371} = 7.14\ (Ans).$$

3.4 ESTIMATIONS OF HALOGEN, SULPHUR AND PHOSPHORUS

These are done by the various methods in the following way:

[A] Accurately weighed sample (very nearly to the macroscale) is taken in a hard glass tube and treated with fuming HNO_3 and heated to 250°C at a safe distance. When the halogens are estimated then small amount of $AgNO_3$ is added to the reaction mixture to prevent the escape of halogen. The products obtained are converted into dilute aqueous solutions and estimated, for halogens by conversions into Ag halide (with $AgNO_3$ soln.), for sulphur by conversion into $BaSO_4$ (with $BaCl_2$ soln.) and for phosphorus by conversion into $Mg_2P_2O_7$ which are filtered, dried and weighed.

The reaction and calculation are given below:

(a) $$\underset{\text{(halogen)}}{\text{Sample } (W_s \text{ gm})} + AgNO_3 + HNO_3 \xrightarrow{\Delta} AgX \xrightarrow[\text{(ii) Dried (iii) Weighed}]{\text{(i) Filtered}} AgX\ (W_X \text{ gm})$$ (estimated).

(i) When X = Cl, then AgX = AgCl.
Mol. wt. of AgCl = 143.5, and x = 35.5 (at. wt.).

$$\text{The \%Cl} = \frac{W_{AgCl} \times 35.5 \times 100}{W_S \times 143.5} \text{ or \% Cl} = \frac{W_{AgCl}}{W_S} \times 24.73 \text{ (Working eqn.)}$$

(ii) When X = Br, then AgX = AgBr
Mol. wt. of AgBr = 188 and x = 80 (at. wt.).

$$\text{The \%Br} = \frac{W_{AgBr} \times 80 \times 100}{W_S \times 188} \text{ or \% Br} = \frac{W_{AgBr}}{W_S} \times 42.55 \text{ (working eqn.)}$$

(iii) When X = I, then AgX = AgI.
Mol. wt. of AgI = 235, and X = 127 (at. wt.).

$$\text{The \%I} = \frac{W_{AgI} \times 127 \times 100}{W_S \times 235} = \frac{W_{AgI}}{W_S} \times 54.04 \text{ (working eqn.).}$$

(b) $$\text{Sample } (W_S \text{ gm}) + HNO_3 \xrightarrow{\Delta} [H_2SO_4] \xrightarrow{BaCl_2} BaSO_4 \text{ (estimated).}$$
Mol. wt. of $BaSO_4$ = 233 and S = 32 (at. wt.).

$$\text{The \%S} = \frac{W_{BaSO_4} \times 32 \times 100}{W_S \times 233} \text{ or } \frac{W_{BaSO_4}}{W_S} \times 13.73 \text{ (working eqn.).}$$

(e) $$\text{Sample } (W_S \text{ gm}) + HNO_3 \rightarrow H_3PO_4 \rightarrow Mg_2P_2O_7 \text{ (estimated).}$$
Mol. wt. of $Mg_2P_2O_7$ = 222 and P = 31 (at. wt.).

$$\text{The \%P} = \frac{W_{Mg_2P_2O_7} \times 62 \times 100}{W_S \times 222} \text{ or } \frac{W_{Mg_2P_2O_7}}{W_S} \times 27.92.$$

Examples:

(i) Wt. of chlorocompound taken = 0.0945 gm (W_S), W_{AgCl} = 0.1435 gm. What is the % of Cl in the compound?

Ans: According to the working eqn. $\%Cl = \frac{W_{AgCl}}{W_S} \times 24.73$ or $\%Cl = \frac{0.1435}{0.0945} \times 24.73$ So the %Cl in the sample is 37.55 (*Ans*).

(ii) Wt. of sulphur compound taken = 0.0172 gm (W_S), $BaSO_4$ = 0.0233 gm. What is the %S in the compound?

Ans: According to the working eqn. $\%S = \frac{W_{BaSO_4}}{W_S} \times 13.73.$

So $\%S = \frac{0.0233}{0.0172} \times 13.73$, or %S = 18.59 (*Ans*).

[B] A new method involves the combustion of the samples in ashless filter paper contained in the "oxygen flask" in the atmosphere of O_2 and ignited electrically. The halogens or SO_2 are absorbed in suitable solvent and estimated in the usual way.

A few more examples are given below (vide: Organic Chemistry by S.P. Banerjee).

(a) $W_S = 0.1435$, $W_{CO_2} = 0.2576$, $W_{H_2O} = 0.0352$. Calculate %C and %H.

Ans: From the working eqn. $\%C = \frac{W_{CO_2}}{W_S} \times 27.27$ or $\frac{0.2576}{0.1435} \times 27.27$

= 48.9 (*Ans*).

$$\%H = \frac{W_{H_2O}}{W_S} \times 11.11 = \frac{0.0352}{0.1435} \times 11.11 = 2.72 \text{ (Ans)}.$$

(b) Estimation of N by Duma, $W_S = 0.2$, Vol. of N_2 = 21.8 ml at 15°C, and (P – f) = 760 mm.

Ans: From the working eqn, $\%N = \frac{V_N(P-f) \times 0.0449}{W_S \times (273+15)} =$

$$\frac{21.8 \times 760 \times 0.0449}{0.2 \times 288} = 12.91 \text{ (Ans)}.$$

(c) Estimation of N by Kjeldahl,
$W_S = 0.2950$, NH_3 neutralises 50 ml of 0.1 (N) H_2SO_4 . What is %N?

Ans: From the working eqn. $\%N = \frac{[x(N) \times V_a - y(N).V_b] \times 1.4}{W_S \times N} = \frac{5 \times 1.4}{0.2950}$

= 23.72 (*Ans*).

(d) Estimation of N by Kjeldahl.

$W_S = 0.4$, NH_3 gas is passed through 25 ml of 1 (N) HCl. After the experiment the reaction mixture (residual acid) is diluted to 150 ml. From this diluted solution 10 ml is neutralised by 7.75 ml of 0.1 (N) NaOH soln. What is %N?

Ans: Since 10 ml of dil. HCl requires 7.75 ml of 0.1 (N) NaOH, the total residual HCl should require $\frac{7.75 \times 150}{10}$ ml of 0.1 (N) NaOH = 11.62 ml of (N) NaOH. From the working eqn,

$$\% N = \frac{[x(N) \times V_a - y(N).V_b] \times 1.4}{W_S \times N} = \frac{(25\,N - 11.62\,N) \times 1.4}{0.4 \times N}$$

$$= \frac{13.38 \times 1.4}{0.4} = 46.83\ (\textit{Ans}).$$

(e) $W_S = 0.25g$, $W_{CO_2} = 0.72g$, $W_{H_2O} = 0.189g$, what are %C and %H?

Ans: From the working eqn. $\%C = \frac{W_{CO_2}}{W_S} \times 27.27 = \frac{0.72}{0.25} \times 27.27$ = 78.53 (*Ans*), and from working eqn. $\%H = \frac{W_{H_2O}}{W_S} \times 11.11 = \frac{0.189}{0.25} \times 11.11$ = 8.39 (*Ans*).

(f) From Kjeldahl's analysis,

$W_S = 0.8$, NH_3 dissolved in 100 ml in 1 (N) H_2SO_4. The residual acid requires 80 ml of 1 (N) NaOH for complete neutralisation. What is %N?

Ans: From the working eqn. $\%N = \frac{[x(N) \times V_a - y(N) \times V_b]}{W_S \times N} \times 1.4$

$$= \frac{[100N - 80N]}{0.8N} \times 1.4 = 35\ (\textit{Ans}).$$

3.5 ESTIMATION OF OXYGEN IN THE SAMPLE

Generally, the %O is calculated by difference, (100 – total of % other elements). This is easy when all the elements have been quantitatively determined. But if necessity arises then the sample is accurately weighed and pyrolysed at high temperature. The gaseous product, thus formed, are driven by pure N_2 over heated carbon (temp. 1120°C) when the entire oxygen present in the sample is quantitatively converted into carbon monoxide. This CO is reacted with iodine pentoxide and the iodine thus liberated is estimated in the usual way which leads to the determination of O in the sample.

3.6 DETERMINATION OF MOLECULAR FORMULA

The aforementioned elemental analyses are carried out with a view to establishing the identities with the known sample or determine the chemical formulas (i.e. molecular formulas). In the latter case, the molecular weight determination becomes essential excepting a few conditions, e.g. when the degree of unsaturation of the derived empirical formula in zero (vide 3.8 (D)]. It is desirable to know the following points before dealing with the molecular weight determination.

(a) Symbols of elements with their gross atomic weights (in parentheses). The symbols marked with * should be committed to memory. These elements are generally more used in organic chemistry.
*H(1); Li(7), *Na(23), *K(39), *Cu(63.5), *Ag(108), *Au(197), *Hg(200); *Mg(24), *Ca(40), Ba(137); B(9); *Al (27), *C(12), *Si(28), Sn(119), Pb(207); *N(14), *P(31), As(75); *O(16), *S(32); Zn(65), *Fe (56), Co(59), Ni(59), *Mn(55), *Pt(195).

(b) When the molecular formula does not correspond with the molecular weight then the term formula weight may be used instead of molecular weight, e.g. in case of water, $(H_2O)_n$, the term molecular weight may be replaced by the term "formula weight" of water is 18.
Molecular weight is the aggregate of the atomic weights of each one of the atoms present in the molecule. Thus if the molecule is $C_nH_mO_P$ then the molecular weight would be n × 12 (at. wt. of C) + m × 1 (at. wt. of H_2) + p × 16 (at. wt. of O_2). The vapour density (D) is half of moleculear weight of the compound when it is in gaseous or vapour state.

(i) General example of calculation of molecular weight: Let the compound be $C_nH_xN_aO_yX_z$. The calculation of molecular weight would be
At. wt. of C × No. of atoms of C = 12 × n (in the present example)

"	H × x	"	H = 1 × x
"	N × a	"	N = 14 × a
"	O × y	"	O = 16 × y
"	X × z	"	X = Q × Z (when X is known, Q will be known).

So, the mol. wt. will be (12n + x + 14a + 16y + Q·z) = M,

and the vapour density of the compound is $\frac{M}{2}$ = D.

(ii) Specific example: Let the mol. formula is,
(a) $C_4H_6Cl_2N_2O$, mol. wt. = (12 × 4 + 1 × 6 + 35.5 × 2 + 14 × 2 + 16 × 1) = 169 = M;
(b) C_2H_5OH, mol. wt. = (24 + 5 + 16 + 1) = 46 = M; (D = 23)

(c) C_5H_5N, mol. wt. $= (60 + 5 + 14) = 79 = M$; $(D = 39.5)$

(d) C_6H_5Cl, mol. wt. $= (72 + 5 + 35.5) = 112.5 = M$; $(D = 56.25)$

It is to be noted that M and D may be even, odd or fractional numbers.

3.7 CALCULATION OF PERCENTAGE OF ELEMENTS IN A COMPOUND

(i) General method: If the compound is $C_nH_xN_aO_y$, then its M (net. wt.) will be $(12 \cdot n + x + 14 \cdot a + 16 \cdot y) = M$.

So the Mgm. of the compound contains $12 \times n$ gm of carbon. Therefore, the percentage of C in the compound is $\frac{12 \cdot n \text{ gm}}{M \text{ gm}} \times 100$

In this way, $\%H = \frac{x}{M} \times 100$, $\%N = \frac{14 \cdot a}{M} \times 100$ and $\%O = \frac{16 \cdot y}{M} \times 100$. It should be noted that $\%C + \%H + \%N + \%O = 100$. Naturally, the $\%O$ is $(100 - \%C - \%H - \%N)$. So the %O is generally obtained through calculation (not by experimental method).

(ii) Specific examples:

(a) Calculate the %C and %H in naphthalene ($C_{10}H_8$).

Ans: The mol. wt. (M) of naphthalene is $(10 \times 12 + 8 \times 1) = 128$.

Therefore, $\%C = \frac{10 \times 12}{128} \times 100 = 93.75$, and $\%H = \frac{8}{128} \times 100 = 6.25$.

$\%C + \%H = (93.75 + 6.25)\% = 100\%$. So the naphthalene does not contain O.

(b) Determine the percentage composition of trichlorobutene.

Ans: The given compound is $C_4H_5Cl_3$ (since this is butene, an alkene). Therefore, the M of the compd. is $(12 \times 4 + 5 + 35.5 \times 3) = 159.5$.

So $\%C = \frac{12 \times 4}{159.5} \times 100 = 30.09$; $\%H = \frac{5}{159.5} \times 100 = 3.13$, $\%Cl = \frac{35.5 \times 3}{159.5}$; 100 = 66.77. Since $(30.09 + 3.13 + 66.77) = 99.99$. The compd. does not contain oxygen.

(c) Find out the percentages of elements in ethanol.

Ans: The formula for ethanol is C_2H_5OH, $= C_2H_6O$.

Therefore, the M is $(2 \times 12 + 6 + 16) = 46$.

So $\%C = \frac{24 \times 100}{46} = 52.17\%$; $\%H = 13.04$; $\%O = \frac{16}{46} \times 100 = 34.78$.

Since $(52.177 + 13.04 + 34.78) = 99.99$, no other element is present.

(d) What are the percentages of the elements in aminopropenol?

Ans: The given compound has formula C_3H_7NO

Therefore, the M of the compound is (36 +7 + 14 + 16) = 73.

So, $\%C = \frac{36}{73} \times 100 = 49.31$; %H = 9.59; %N = 19.19.

The total percentage of C, H and N is 49.31 + 9.59 + 19.19 = 78.09.

So, by difference %O is (100 – 78.09) = 21.91.

Or, by calculation %O is $\frac{16}{73} \times 100 = 21.91$.

3.8 DETERMINATION OF ELEMENTAL FORMULA AND EMPIRICAL FORMULA

(i) From the percentages of the elements present in the sample its elemental formula (i.e the elements which make the compound) is obtained.

(ii) The empirical formula of the compound is calculated step by step (arithmetical formula) as shown in the following

(iii) The chemical formula (exact molecular formula) is derived from the molecular weight and the calculated empirical formula. In some cases, the empirical formula becomes equal to chemical formula thus eliminating the use of molecular weight determination.

(iv) In $C_nH_mO_xS_y$ the chemical formula (i.e. final formula) the H must be in whole number and will never cross the limit of 2n + 2 (for C_n-compounds) or its equivalent (vide ... 3.10,b.).

(v) **Specific examples:** From the %C = 82.75 and %H = 17.25, the molecular formula is to be derived.

Ans: *First step:* From the given data (82.75 + 17.25) = 100 is obtained. So no other element is present. So the *elemental formula* of the compound is CH.

Second step: The simple ratio of the elements in the compound is determined by dividing the percentage of the element by its own atomic weight. If the result is fractional atom then it should be multiplied by suitable factors or divided again by the smallest fractions to arrive at the nearest whole number of all the atoms in the molecule. Thus, the calculated (i.e. empirical formula) of the compound is $C_{\frac{82.75}{12}} H_{\frac{17.25}{1}}$ or $C_{6.8}H_{17.25}$. Dividing by the smallest number (6.8 here), $C_{\frac{6.8}{6.8}} H_{\frac{17.25}{6.8}}$ or = $C_1H_{2.5}$. But H cannot be fractional in any compound. So this is multiplied by 2 (suitable factor to make H as well as the other elements whole number). So the *empirical formula* is C_2H_5 of the compound.

Third step: Since the compound is a hydrocarbon, the number of H in the molecule cannot be odd (5 in the present case). Hence, the empirical formula is multiplied by 2 to arrive at C_4H_{10}. In this formula, the number of H is maximum (reference C_nH_{2n+2} for alkanes). Hence this must be the desired molecular formula of the compound. It is to be noted that the *degree of unsataration* (DU) [vide 3.10.] is zero and molecular weight determination or the vapour density is not essential to derive the molecular formula here.

(iv) *Summary* of the procedure to calculate the molecular formula from elemental analyses data of the present example:

(a) $C_{82.75} + H_{17.25} = 100$ (So no other element is present). So *elemental form* = CH.

(b) $C_{\frac{82.75}{12}}H_{\frac{17.25}{1}} = C_{6.8}H_{17.25}$ (cannot be any formula due to fractional atoms).

(c) $C_{\frac{6.8}{6.8}}H_{\frac{17.25}{6.8}} = C_1H_{2.5}$ (divided by smallest quotient to get an atom whole numbered.)

(d) $(C_1H_{2.5}) \times 2 = C_2H_5$ (all atoms are whole numbered but H is odd; not molecular formula).

(e) $(C_2H_5) \times 2 = C_4H_{10}$

This is the molecular formula because the number of H is maximum ($C_{2n} + 2 = 10, n = 4$), which is equivalent to degree of unsaturation zero in the present example. The molecular weight is not necessary in this case.

(vii) **Other examples:**

Example A: The compound contains C, 92.30% and H 7.69%. Calculate the mol. formula.

Ans:

(a) Since 92.30 + 7.69 = 99.99 (no elements other than C, H is present. So the *elemental formula* is CH).

(b) $C_{\frac{92.30}{12}}H_{\frac{7.69}{1}} = C_{7.69}H_{7.69}$ (dividing by respective at. wt. of elements)

(c) $C_{\frac{7.69}{7.69}}H_{\frac{7.69}{7.69}} = CH$ (this is *empirical formula* because the atoms are in whole number ratio), but this is not molecular formula (M) because H is odd number).

(d) $(CH)_2$, $(CH)_4$, = $(CH)_m$ = (These are the *molecular formula*, m must be even number leading to H_m even number)

(e) Therefore C_2H_2, C_4H_4, C_6H_6 etc. are answers.

(f) The exact mol. formula (i.e. chemical formula) can be determined if either vapour density (D) or molecular weight (M) is given. Thus, when D = 39, the M is 78 (i.e. 2D) and $(CH)_m$ = 78 or m × 13 = 78, so m = 6. So, the molecular formula of the compound must be C_6H_6 (*Ans*).

Example B: The compound contains C, 46.6%; H, 7.7% and Cl, 45.8%, Find out the molecular formula of the compound.

Ans:

(a) 46.4 (C) + 7.7 (H) + 45.8 (Cl) = 99.9 (the 0.1% is ignored). So the *elemental formula* is CHCl.

(b) $C_{\frac{46.4}{12}}H_{\frac{7.7}{1}}Cl_{\frac{45.8}{35.5}} = C_{3.86}H_{7.7}Cl_{1.29}$ (Ratio of the atoms of elements).

(c) $C_{\frac{3.86}{1.29}}H_{\frac{7.7}{1.29}}Cl_{\frac{1.29}{1.29}} = C_{2.99}H_{5.96}Cl_1$ or C_3H_6Cl (going to nearest whole number). This is *empirical formula*.

(d) The D or M is not given, so in $(C_3H_6Cl)_m$ = M, the m cannot be known.

(e) The empirical formula is multiplied by 2. So $(C_3H_6Cl) \times 2 = C_6H_{12}Cl_2$.

(f) The degree of unsaturation is C_3H_6Cl is + 0.5 (fractional). So this is not the molecular formula. But in $C_6H_{12}Cl_2$ the degree of unsataration is zero. Hence this is the molecular formula of the given compound (M = 155). But the problem has been solved in spite of the absence of molecular weight.

(g) Suppose the multiplying factor m (in d above) is more than 2, then the degree of unsaturation of the compound would be either fractional or negative which are inadmissible for any compound. Thus, $(C_3H_6Cl)_3 = C_9H_{18}Cl_{13}$, degree of unsaturation is –0.5, in $(C_3H_6Cl)_4$, i.e. the degree of unsaturation in $C_{12}H_{24}Cl_4$, is –1.

(h) Thus $C_4H_{10}O$ does not require the molecular weight, because it is the molecular formula. Similarly, $C_3H_7O_3$ does require the molecular weight because this is empirical formula surely, but not the molecular formula. But on multiplying by 2 (i.e. m), the formula becomes $(C_3H_7O_3)_2 = C_6H_{14}O_6$ which has the maximum number of H atoms admissible in the compound (DU = degree of unsaturation = 0),

Example C: What is the molecular formula of the compound having the following analytical data: %C, 44.05; %H, 7.31; %Cl 26.19 and %N, 10.18?

[It is to be noted that the molecular weight has not been given, yet the molecular formula has been wanted.]

Ans: Total % gives (44.05 + 7.31 + 26.19 + 10.18) = 87.73. So %O is 12.27.

(a) Therefore, the *elemental formula* of the compound is CHCl NO.

(b) Ratio of elements, $C_{\frac{44.05}{12}}H_{\frac{7.31}{1}}Cl_{\frac{26.19}{35.5}}N_{\frac{10.18}{14}}O_{\frac{12.27}{16}} = C_{3.67}H_{7.31}Cl_{0.74}$ $N_{0.73}O_{0.76}$

(c) Dividing each one by the smallest quotient, $C_{\frac{3.67}{0.73}}H_{\frac{7.31}{0.73}}Cl_{\frac{0.74}{0.73}}N_{\frac{0.73}{0.73}}O_{\frac{0.76}{0.73}}$ $= C_{5.02}H_{10.0}Cl_{1.01}N_{1}O_{1.04}$.

(d) The quotients are taken to the nearest whole number = $C_5H_{10}ClNO$, this is the *empirical formula* of the compound.

(e) The molecular weight (M) has not been given, the empirical formula, obtained in (d) can be molecular formula since its degree of unsaturation (DU) is 1 (of the empirical formula). Similary, $(C_5H_{10}ClNO)x$ (x = whole number) can be the possibities, since all of them have Du. = 1. Therefore M is needed to deliver exact answer.

Example D: The percentage composition of the compound is, C, 43.63; H, 8.72; Cl, 25.8, N, 10.18. What is the molecular formula?

Ans: Total = 43.63 + 8.72 + 25.8 + 10.18 = 88.33, So %O = 11.67.

So the *elemental formula* is CHCl NO.

The *empirical formula* is $C_{\frac{43.63}{12}}H_{\frac{8.72}{1}}Cl_{\frac{25.8}{35.5}}N_{\frac{10.18}{14}}O_{\frac{11.67}{16}} = C_{3.63}H_{8.72}Cl_{0.72}$ $N_{0.72}O_{0.72}$, dividing by 0.72, the formula becomes = $C_5H_{12}ClNO$.

This should be the *molecular formula* since this is completely saturated, or in other words, the DU is zero. So $C_5H_{12}ClNO$ (*Ans*).

So it is evident that when the Du. of emperical formula be zero then it is an unnecessary to determine the molecular formula.

3.9 CALCULATION OF MOLECULAR FORMULA WHEN THE MOLECULAR WEIGHT IS KNOWN

It has been shown in **3.7** that the % of C in a compound (or an element E in the compound) is given by the following expression, if the M is known.

$$\%C = \frac{\text{at. wt}_C \times NO_C \times 100}{M}, \text{ general working eqn. is } \%E = \frac{\text{at. wt}_E \times NO_E \times 100}{M}$$

So $NO_E = \dfrac{\%E \times M}{\text{at. wt}_E \times 100}$, where NO_E is the number of atoms of element E in the molecule of the compound having molecular weight M, and at. wt_E is the atomic weight of E. Therefore, when E = Carbon, the eqn. is $NO_C = \dfrac{\%C \times M}{12 \times 100}$.

Example A: In the compound %C, 43.79; %H, 4.37; %Cl, 51.82; and M = 548: What is the molecular formula?

Ans:

(i) The total % is (43.79 + 4.37 + 51.82) = 99.98. So oxygen is absent.

(ii) From the working eqn. $NO_C = \dfrac{43.79 \times 548}{1200} = 19.99 = 20.$

$$NO_H = \frac{4.37}{100} \times 548 = 23.94 = 24$$

$$NO_{Cl} = \frac{51.82 \times 548}{35.5 \times 100} = 7.99 = 8.$$

Therefore the *molecular formula* of the compound is $C_{20}H_{24}Cl_8$ (*Ans*).

Example B: The compound contains, C, 33.2%, H, 4.6%; Cl, 32.7%

Mol. wt. is 108.5. What is the molecular formula?

Ans:

(i) The total is (33.2 + 4.6 + 32.7) = 70.5%, So %O is (100 – 70.5) = 29.5%
Therefore, the *elements formula* is CHClO.

(ii) The *molecular formula* is calculated using the working formula,

$$No_E = \frac{\% E \times M}{\text{at. wt}_E \times 100} \text{ or } NO_C = \frac{33.2}{1200} \times 108.5 = 3.00 = 3$$

$$NO_H = \frac{46}{100} \times 108.5 = 4.99 = 5$$

$$NO_{Cl} = \frac{32.7 \times 108.5}{35.5 \times 100} = 0.999 = 1$$

$$\text{and } NO_O = \frac{29.5 \times 108.5}{1600} = 2.00 = 2$$

so the *molecular formula* is $C_3H_5ClO_2$ (*Ans*)

Example C: The analyses are, C, 35.3%; H, 5.3%; Cl, 41.7%; N, 8.2%; O, 9.4%. The mol. wt. (M) is 170. What is the mol. formula of the compound analysed ?

Ans:

(i) Total composition (35.3 + 5.3 + 41.7 + 8.2 + 9.4) = 99.9.

(ii) The mol. formula is

$$NO_C \frac{35.3}{1200} \times 170, NO_H \frac{5.3}{100} \times 170, NO_{Cl} \frac{41.7}{3550} \times 170, NO_N \frac{8.7}{1400} \times 170,$$

$$NO_O \frac{9.4}{1600} \times 170$$

or $C_5H_9Cl_2N_1O_1$ (*Ans*).

Example D: The compound contains 50% C and mol. wt. is 192. What is the molecular formula of the compound?

Ans:

(i) Total C = 50 % (so O = 100 – 50) = 50%. The compound contains 50% C and 50% O.

(ii) From the working formula, $\%C = \frac{at.wt_C \times NO_C \times 100}{M}$ or $50 = \frac{12 \times NO_C \times 100}{M}$ or $\frac{M \times 50}{1200} = NO_C$. Therefore $\frac{192 \times 50}{1200} = NO_C = C_8$, and $\frac{50 \times 192}{1600} = NO_O = O_6$, so the formula is C_8O_6 (*Ans*).

3.10 DOUBLE BOND EQUIVALENT—ALSO CALLED DEGREE OF UNSATURATION OR UNSATURATION NUMBER

From the foregoing discussions it is evident that degree of unsaturation (D.U.), often called unsaturation number (U.N.) or double bond equivalent (D.E.), plays a very important part in determining molecular formula as well as structure elucidation of organic compounds, with special reference to the natural products. So the D.U. is explained in the following in more detail.

(a) Alkane C_nH_{2n+2}[A] (Saturated) $\xrightarrow{-2H}$ Alkene C_nH_{2n}[B] (Unsaturated) one double bond. $\xrightarrow{-2H}$ Alkyne C_nH_{2n-2}[C] (Unsaturated) one triple bond.)

Alkene [B] $\xrightarrow{-2H}$ Alkadiene C_nH_{2n-2}[D] (Unsaturated) two double bonds.

Alkane [A] $\xrightarrow{-2H}$ Cycloalkane C_nH_{2n}[E] (Saturated)

Cycloalkane [E] $\xrightarrow{-2H}$ C_nH_{2n-2}[F] (Unsaturated) cyclic with one double bond

It is observed that [B] and [E] are isomeric and [C], [D] and [F] are also isomeric because they have the same molecular formula. This shows that a cycle or a double bond is introduced into the parent compound by the loss of 2 H atoms at the same time.

(b) The degree of unsaturation (D.U.) is the number of double bond or cycle (or a combination of both). The triple bond has DU 2 (equivalent to two double bonds). The D.U. is calculated in the following way:

(i) In the compound containing C_nH_x, D.U. $= (n+1) - \left(\frac{x}{2}\right) = \left(\frac{2n+2-x}{2}\right)$.

(ii) In the compound containing $C_nH_xO_y$, D.U. $= (n+1) - \left(\frac{x}{2}\right)$, Oxygen (or any bivalent element) is immaterial.

(iii) In the compound containing $C_nH_xX_y$, D.U. = $(n + 1) - \left(\frac{x+y}{2}\right)$, X ≡ H (any univalent element).

(iv) In the compound containing $C_nH_xN_a$, D.U. = $(n + 1) - \left(\frac{x-a}{2}\right)$, N ≡ (any trivalent element).

(v) The compound containing $C_n\,H_x\,X_z\,N_a\,O_y\,S_b$ (x = halogen)

D.U. = $(n + 1) - \left(\frac{x+z-a}{2}\right)$ (y and b are ignored as O and S are divalent. elements).

(vi) The compound containing $C_nH_xCl_zBr_dI_eN_aO_yS_b$, D.U. = $(n + 1) - \left(\frac{x+z+d+e-a}{2}\right)$.

(e) **Example:** Only one example in each case is cited below (with C_4 hydrocarbon).

(i) C_nH_x (n = 4, x = 10), D.U. $C_4H_{10} = (4+1) - \left(\frac{10}{2}\right) = 5 - 5 = 0$, so compd. is alkane (saturated).

(ii) C_nH_x (n = 4, x = 8), D.U. $C_4H_8 = (4 + 1) - \left(\frac{8}{2}\right) = 5 - 4$

$$= \begin{cases} 1 \text{ so compd. is alkene (1 double bond).} \\ 1 \text{ so compd. is cycloalkane (1 ring)} \end{cases}$$

Therefore , 1 D.U. ≡ 1 [C = C] or 1 cycle.

(iii) C_nH_x (n = 4, x = 6) D.U. = $C_4H_6 = (4 + 1) - \left(\frac{6}{2}\right) = 5 - 3 = 2$, So compd. (diene or dicyclic or a cycle + a double bond, or alkyne).

(iv) C_nH_x, (n = 4, x = 4) D.U. = $C_4\,H_4 = (4 + 1) - \left(\frac{4}{2}\right) = 5 - 2 = 3$ (combination of C = C, cycle and C ≡ C).

(v) C_nH_x (n = 4, x = 2), D.U. = $C_4H_2 = (4 + 1) - \frac{2}{2} = 5 - 1 = 4$ (-do-).

(vi) C_nH_x (n = 4, x = 0), D.U. = $C_4O_y = (4 + 1) - 0 = 5$ = At least theoretically possible.

Examples:

Thus,

C_4H_{10}, 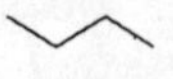 or 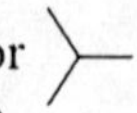 C_4H_8 Du = 1. 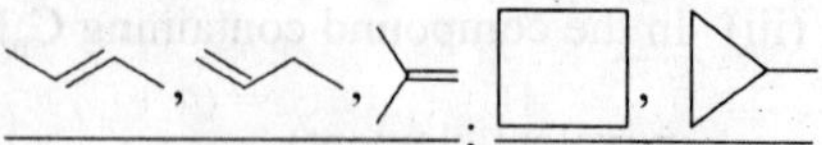

Du = 0, (alkane) C = C Cycle

C_4H_6, 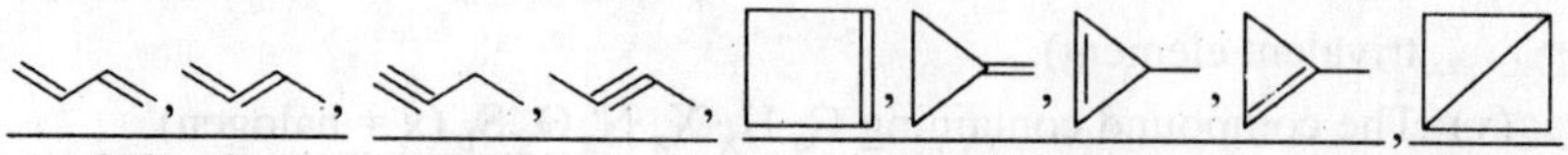

Du = 2, 2(C = C) 1(C ≡ C) 1 cycle + 1 (C = C) 2 Cycles

C_4H_4, $H_2C = C = C = CH_2$, $HC \equiv C - CH = CH_2$, 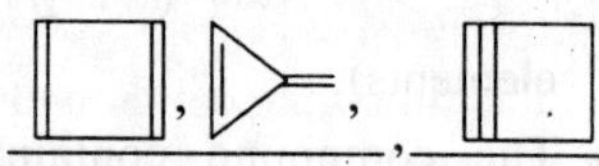

Du = 3, 3(C = C) 1(C = C) + 1 (C ≡ C) 1 cycle + 2(C = C) Theoretical

C_4H_2, $HC \equiv C - C \equiv CH$, 

Du = 4, 2(C ≡ C)

Theoretical

C_4O, D = 5, C_4O is 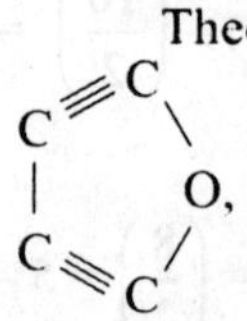 C_4O_2, 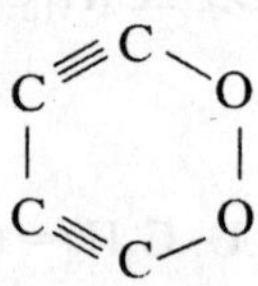 Note, oxygen is immaterial, its number is ignored.

Theoretical Theoretical

(d) The following points are to be noted:

(i) If the molecule contains an n-carbon atoms then the miximum number of H atoms (or any other monovalent atoms) will be 2n + 2 (This is x in C_nH_x). When the number of H exceeds this (2n + 2) limit, the compound will be non-existent. If the compound contains C, H and halogens, then number of H + number of halogen must not exceed (2n + 2) in $C_nH_xX_z$. Moreover the aggregate of H atoms and halogen atoms must be always even number. Again the D.U. will never be fractional or negative.

Examples:

(H + Hal) = even, D.U. = 0

$C_4H_5Cl_5$, D.U. $= (4+1) - \left(\frac{5+5}{2}\right) = 5 - 5 = 0$, saturated compd., it may exist.

$C_4H_2Cl_2Br_4I_2$,; – D.U. $= 5 - \left(\frac{2+2+4+2}{2}\right) = 5 - \left(\frac{10}{2}\right) = 0$, -do-

$C_3H_5Cl_3$; D.U. $= 4 - \left(\frac{5+3}{2}\right) = 4 - 4 = 0$, -do-

$C_6H_9ClBrIF_2$; D.U. $= 7 - \left(\frac{9+1+1+1+2}{2}\right) = 7 - \left(\frac{14}{2}\right) = 0$, -do-

(ii) The following compounds do not exist.

(H + Hal) = odd also D.U. is fractional

$C_4H_5Cl_4$; D.U. $= (4 + 1) - \left(\frac{5+4}{2}\right) = 5 - 4.5 = 0.5$. Does not exist (sure).

$C_4H_2Cl_2Br_3I_2$; D.U. $= (4 + 1) - \left(\frac{2+2+3+2}{2}\right) = 5 - 4.5, = 0.5$ -do-

$C_6H_6ClBrIF_2$; D.U. $= (6 + 1) - \left(\frac{6+1+1+1+2}{2}\right) = 7 - \left(\frac{11}{2}\right) = 7 - 5.5$ $= 1.5$, -do-

(iii) The following compounds do not exist because (H + Hal = more than 2n + 2),

C_3H_{10}, D.U. $= (3 + 1) - \left(\frac{10}{2}\right) = 4 - 5 = -1$, does not exist (sure).

$C_4H_2Cl_5Br_5$, D.U. $= (4 + 1) - \left(\frac{2+5+5}{2}\right) = 5 - \frac{12}{2} = -1$, -do-

$C_6H_9Cl_4Br_3$; D.U. $= 7 - \left(\frac{9+4+3}{2}\right) = 7 - 8 = -1$, -do-

(iv) The divalent elements do not influence the D.U. They are immaterial.

$C_3H_8O_2$, D.U. $= (3+1) - \left(\frac{8}{2}\right) = 4 - 4 = 0$, This compd. exists, this is saturated.

$C_4H_8O_3$, D.U. $= (4 + 1) - \left(\frac{8}{2}\right) = 5 - 4 = 1$, -do-, this is olefinic or cyclic.

$C_5H_6O_3S$, D.U. $= (5 + 1) - \left(\frac{6}{2}\right) = 6 - 3 = 3$, -do-, various unsaturated and cyclic compounds.

(v) When the compounds contain trivalent element(s) like N, P, etc. then the sum of atoms of H and atom(s) of trivalent element(s) must be always even and H will never exceed [2n + 2 + no. of trivalent heteroatom(s)], examples are

C_3H_9N, D.U. $= (3 + 1) - \left(\frac{9-1}{2}\right) = 4 - 4 = 0$, sum of H + N = even (9 + 1 = 10).

Likewise, CHN, $C_2H_4N_2$, C_3H_7N, $C_4H_5N_3$, $C_5H_{12}P_2$ may exist.

But, $C_2H_3N_2$, C_3H_8N, $C_4H_{14}N_2$, $C_5H_{16}P_2$ never exist.

Calculation of D.U. = $C_nH_xHet_y = (n + 1) - \left(\frac{x - y}{2}\right)$ = if positive whole number or zero, the compd. may exist but if negative or fractional, the compounds never exist.

Examples:

(i) $C_4H_{15}N_3$, D.U. $= (4 + 1) - \left(\frac{15-3}{2}\right) = 5 - \frac{12}{2} = -1$.

This compd. does not exist although sum of H and N is even.

(ii) $C_6H_8N_3P_2$, D.U. $= (6 + 1) - \left(\frac{8-3-2}{2}\right) = 7 - \frac{3}{2} = 7 - 1.5 = 5.5$. The compd. is non-existent.

3.11 SUMMARY OF DEGREE OF UNSATURATION CALCULATION AND INFERENCES

(i) The C_nH_x representing alkane, alkene, alkyne and alicyclic branched or unbranched hydrocarbons will always have mol. wt. even, x must be even number and never more than 2n + 2.

(ii) In $C_nH_xA_y$ (A = divalent element like O, S, etc.), the A_y will be ignored, and as in (i) x = even number and not more than 2n + 2.

(iii) In $C_nH_xX_z$ (X = halogen or univalent element), the sum of (x + z) will be even number and never more than (2n + 2). During calculation of D.U. the halogen atom will be considered as H atom.

(iv) In $C_nH_xE_a$ (E = trivalent element like N, P, etc.), the sum of (x + a) will be even number, and x will never exceed (2n + 2 + a, maximum number of H).

(v) In $C_nH_xX_zA_yE_a$ (X = univalent, A = bivalent, E = trivalent elements), the sum of (x + z + a) will be even member, y will be ignored, and (x + z) will never exceed (2n + 2 + a). The D.U. (i.e. U.N or DBE) of such molecules will be = $(n + 1) - \left(\frac{x+z-a}{2}\right)$.

(vi) The D.U. will be either zero or positive whole number but never fractional or negative whole number. If during calculation of molecular formula the D.U. is checked and found to be negative or fractional value, then the calculation or the given data is wrong. The result needs to be multiplied by suitable factor to satisfy the conditions laid in [(i) to (v) above] when the DU is fractional (positive).

(vii) $C_nH_xN_aO_y$ requires (x + a) = even number, y = immaterial, maximum value for x = (2n + 2 + a), and D.U. = $(n + 1) - \left(\frac{x-a}{2}\right)$.

(viii) The D.U. is not restricted to C – C bonds but it may be extended to C = O or C = N or C ≡ N, i.e. between C – heteroatom also.

(ix) When the D.U. = O, the compound is open chain and saturated.

Conclusion

The calculation of D.U. helps to check up the correctness of mol. formula and in problem solving exercise.

Examples:

(i) C_4H_9, C_4H_7, C_2H_3, etc. are not compounds as the no. of H is odd and the D.U. is fractional.

(ii) C_4H_{14}, C_2H_8, C_8H_{20}, etc. are not compounds as the no. of H > 2n + 2.

(iii) $C_4H_5O_2$, $C_5H_{11}S$, C_2H_5O, etc. are not compounds as no. of H is not even number.

(iv) $C_4H_{12}O$, $C_2H_{14}OS$, etc. are not compounds as no. of H > 2n + 2.

(v) $C_3H_3N_2$, C_4H_8N, $C_5H_{12}N_3$ are not compounds, as no. of H + no. of N is not even number.

(vi) $C_3H_{11}N_2$, $C_4H_{14}N_2$, $C_5H_{20}N_4$, etc. are not compounds as no. of H > (2n + 2 + no. of N).

(vii) C_3H_4Br, $C_4H_7Cl_2$, $C_2H_3Cl_2$, etc. are not compounds as no. of H + no. of halogen is odd.

(viii) $C_4H_7Br_5$, $C_3H_8Cl_2$, $C_2H_6I_2$, etc. are not compounds as no. of H + no. of halogen > (2n + 2).

(ix) C_4H_7N, D.U. = $(4 + 1) - \left(\frac{7-1}{2}\right) = 5 - 3 = 2$. The compound *may* exist.

(x) $C_8H_{14}Cl\,NO$ (in English alphabetic order), D.U. = $(8 + 1) - \left(\frac{14+1-1}{2}\right) = 9 - 7 = 2$, compd. may exist.

(xi) $C_8H_{13}Cl\,N_2S$, D.U. = $(8 + 1) - \left(\frac{13+1-2}{2}\right) = (9 - 6) = 3$, the compound may exist.

(xii) $C_6H_{11}BrN_2\,OS$, D.U. = $(6 + 1) - \left(\frac{11+1-2}{2}\right) = (7 - 5) = 2$, the compound may exist.

(a) It is to be noted that the non-existence of a compound with the given molecular formula can be detected instantaneously with the help of aforementioned rules of D.U., but the existence of the compound cannot be said with certainty even when the D.U. is suggestive of existence.

e.g. $C_6H_{10}BrClN_2S$, D.U. $= (6+1) - \left(\frac{10+1+1-2}{2}\right) = 7-5 = 2$, Such compound *may* exist (not sure) but $C_6H_9BrClN_2S$, D.U. $= (6+1) - \left(\frac{9+1+1-2}{2}\right) = 7-4.5 = 2.5$, such compound does not exist (surely, no doubt).

e.g. $C_5H_{14}NOPS$, D.U. $= 6 - \left(\frac{14-1-1}{2}\right) = 6-6 = 0$, the compound *may* exist (not sure).

e.g. $C_7H_{15}BrCl_2NPSO$; D.U. $= 8 - \left(\frac{15+1+2-1-1}{2}\right) = 8-8 = 0$, compound *possible*, but $C_7H_{15}Br_2Cl_2NPSO$, D.U. $= 8 - \left(\frac{15+2+2-1-1}{2}\right) = 8 - 8.5 = -0.5$, compd. impossible (certain).

(b) Another point may be helpful for quick recognition of the existence of the compound. In $C_nH_xX_z$ if $(x+z)$ is odd, and/or $x > (2n+2-z)$ the compound *does not* exist. Similarly, in $C_nH_xN_a$ if $(x+a)$ is odd, in $C_nH_xX_zN_aO_y$ if $(x+z+a)$ is odd, then these compounds do not exist. In the last compound the value of x will never exceed $(2n+2+a-z)$.

Examples:

(a) The compound is $C_6H_{10}BrClN_2S$.

(i) Since [10(H) + 1(Br) + 1(Cl) + 2(N)] = 14 even no., the compound may exist.

(ii) Again, number of H = 10 < 12(H) + 2 − 1(Cl) − 1(Br) + 2(N), so the compd. may exist.

(b) In $C_nH_kBr_hCl_gI_iN_eO_xP_fS_y$, the following points are noted.

(i) In this compound the *maximum* number of H (m) will be $= (2n+2+e+f-g-h-i)$ and the compound is open chain and saturated.

(ii) Again if $\left(\frac{m-k}{2}\right)$ is fractional or negative (D.U.), the compound does not exist. But *k may be either even or odd* (i.e. number of H given).

(a) $C_5H_{14}NOPS$ is the proposed formula.

Here $H_{max} = m = [(5\times 2) + 2 + 1(N) + 1(P)] = 14$ (*even*)

Given value of H is 14 (= k). Therefore $\left(\frac{m-k}{2}\right) = \left(\frac{14-14}{2}\right) = 0 =$ D.U.

So the formula is possible for an open chain saturated compound.

(b) $C_7H_{15}BrCl_2NPS$ is the proposed formula.
Here $H_{max} = (m) = [7 \times 2 + 2 - 1(Br) - 2(Cl) + 1(N) + 1(P)] = 15$ (*odd*).

Given value of k is 15. Therefore, $\left(\frac{m-k}{2}\right) = \frac{15-15}{2} = 0$. The compound may exist.

(c) $C_7H_{17}BrCl_2NPS$ is the formula.
$H_{max} = [(7 \times 2 + 2 - 1(Br) - 2(Cl) + 1(N) + 1(P)] = 15$ (*odd*) = m.

Given value of k is 17. Therefore, $\left(\frac{m-k}{2}\right) = \frac{15-17}{2} = -1$ (negative).

Such compound does not exist (certain).

(d) $C_7H_{16}Br_2ClNPS$ is the formula, the $H_{max} = [(7 \times 2) + 2 - 2(Br) - 1(Cl) + 1(N) + 1(P) = m = 15$ (*odd*).

Given value of k (i.e. actual no. of H given in the formula) is 16.

Therefore, $\left(\frac{m-k}{2}\right) = \frac{15-16}{2} = -0.5$ (fractional). The compd. does not exist.

3.12 CALCULATION OF H IN A MOLECULE FROM DEGREE OF UNSATURATION

For each D.U. (i.e. unsaturation number or double bond equivalent) two H atoms are to be subtracted from the corresponding open chain saturated compound. Thus,

(i) In $\underset{D_U = 2}{C_nH_{2n-2}} \xrightarrow{[+2H]} \underset{D_U = 1}{C_nH_{2n}} \xrightarrow{[+2H]} \underset{D_U = 0}{C_nH_{2n+2}}$ (The starting compound is alkyne)

(ii) $\underset{D_U = 2}{C_nH_{2n-2}} \xrightarrow{[+2H]} \underset{D_U = 1}{C_nH_{2n}} \xrightarrow{[+2H]} \underset{D_U = 1}{C_nH_{2n}}$ (The starting compound is Cycloalkene)

(iii) $\underset{D_U = 2}{C_nH_{2n-2}} \xrightarrow{[2H]} \underset{D_U = 2}{C_nH_{2n-2}}$ (The starting compound is bicyclic)

Examples (i): 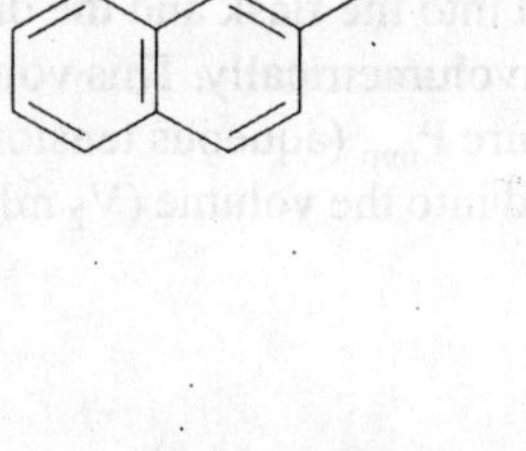 $= C_{17}H_?O$, *Calculation of H*:

D.U. = 4 (cycles) + 9 (double bonds) = 13, so H subtracted ($13 \times 2 = 26$) from the open chain saturated hydrocarbon ($17 \times 2 + 2 - 26$) = 10.

So the given compound $C_{17}H_{10}O$.

Example (ii): Calculation of H in compound [A]

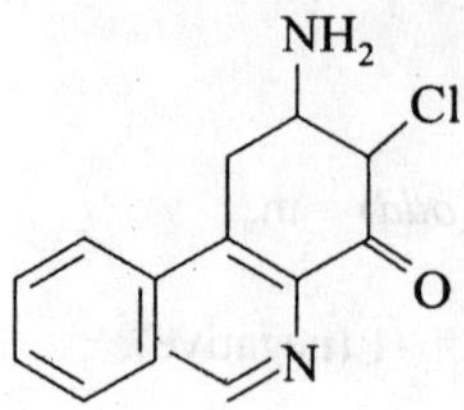

[A]

$= C_{13}H_xClN_2O$
Formula of [A]

The open chain saturated compound will require $[(13 \times 2) + 2 - 1(Cl) + 2(N)] = 29$ H atoms.

The given compound has D.U. = 3 (cycles) + 6 (double bonds) = 9 ≡ 18 H atom. So x is the *calculated* H in [A] = 29 − 18 = 11 H. So the formula for [A] is $C_{13}H_{11}ClN_2O$.

Example (iii):

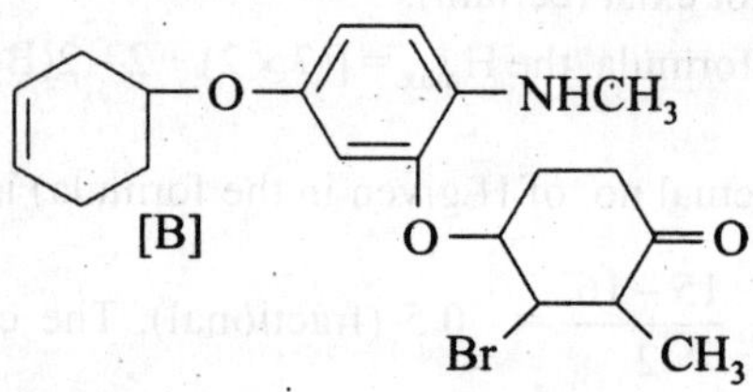

[B]

This is $C_{20}H_xBrNO_3$.

If the compound [B] were open chain saturated compound, the number of H atoms would have been $[(20 \times 2) + 2 - 1(Br) + 1(N)] = 42$.

The D.U. is 3 (cycles) + 5 (double bonds) = 8, ≡ 16 H atoms.

Therefore x = (42 − 16) = 26. So the formula [of B] should be $C_{20}H_{26}BrNO_3$. The *desirability of counting* the H atoms as well as *calculating* the H atoms is to confirm the correctness of the proposed or given formula.

3.13 MOLECULAR WEIGHT DETERMINATION

[A] Physical methods are (i) Hoffmann and Victor Meyer method, (ii) Cryoscopic method, and (iii) Ebullioscopic method.

[B] Chemical methods are (i) Equivalent weight determination, (ii) Analysis of silver salts of carboxylic acids, and (iii) Analysis of chloroplatinate salts of basic compounds.

[C] Instrumental method involves mainly mass spectrometric analysis.

[A] Physical Methods

(i). In the Victor Meyer process, the organic compound (it must not decompose on vaporisation) is accurately weighed. The Victor Meyer flask (two-necked long stem flask, one neck is closed with a stopper) is heated to remove the air inside the flask. The sample (Wgm. = 0 · xyza, x ≈ 2) is introduced into the flask and the displaced residual air (by the vapour of the sample) is measured volumetrically. This volume (V_1 ml), the room temperature t°C and atmospheric pressure P_{mm} (aqueous tension at t°C is f mm) are noted, and the volume (V_1 ml) is converted into the volume (V_2 ml) at N.T. P. Thus,

$$V_2\,\text{ml} = \frac{(P - f)_{mm} \times V_1\,\text{ml} \times 273\,K}{760_{mm} \times (273 + t)\,K}$$ (This is the volume of the vapour from Wgm at NTP).

(a) Direct calculation: (Found, V_2 ml of vapour weighes Wgm) at N.T.P.

Therefore, 22.4 L of vapour weighes $\frac{22400\,\text{ml} \times \text{Wgm}}{V_2\,\text{ml}}$ = Mgm.

Therefore, the molecular weight is M.

Example: W = 0.4250 gm, V_1 = 66 ml, t = 14°C, P = 745 mm ($f^{14°C}$ = 12.3 mm).

In this case, the mol. wt. = M =

$$\frac{22400\,\text{ml} \times 760\,\text{mm} \times (273 + 14)\,K \times 0.4250\,\text{gm}}{(745 - 12.3)\,\text{mm} \times 66\,\text{ml} \times 273\,K} = 157.3\,\text{gm}$$

So the molecular weight = 157.

(b) *via Vapour Density:* 1 ml of H at NTP weighs 0.00009 gm,

So V_2 ml of H at,, 0.00009 gm × V_2.

So the density of vapour is $\frac{\text{Wgm}}{V_2 \times 0.00009\,\text{gm}}$ = D.

So the mol. wt. is 2D = M.

[B] Chemical methods

(a) *Determination of equivalent weight:* The molecular weights of acids and bases are determined by titration with standard solution of inorganic bases and acids, respectively. Thus,

Molecular weight (M) of organic acids = Its equivalent weight (E) × basicity.

Molecular weight (M) of organic base = Its equivalent weight (E) × acidity.

(b) *Determination of molecular weights of organic acid through silver salt method:* The silver salt of an acid is purified, dried and accurately weighed. On strong heating, it decomposes into metallic silver which is weighed to determine the molecular weight of the silver salt ($M_{Ag\,salt}$). From this data the weight of silver (at. wt. 108) is subtracted and weight of H is added to get the molecular weight of the organic acid (M_{acid}).

Thus, if (i) weight of silver salt = $W_{Ag\,salt}$, (ii) metallic silver obtained = W_{Ag}, (iii) the basicity of the organic acid = B, (iv) the molecualr weight of organic acid = M_{acid}, and (v) the molecular weight of the silver salt = $M_{Ag\,salt}$, then

$$M_{Ag\,salt} = \frac{W_{Ag\,salt} \times B \times 108}{W_{Ag}}$$ and $M_{acid} = M_{Ag\,salt} - (B \times 108) + B$.

Examples:

(i) Given $W_{Ag\,salt}$ = 0.70 gm, W_{Ag} = 0.50 gm, B = 2(basicity), M_{acid} = ?

Ans: Since the working equation is $M_{Ag\ salt} - (B \times 108) + B = M_{acid}$, so M_{acid}

$$= \frac{0.70\ gm \times 2 \times 108}{0.50\ gm} - (2 \times 108) + 2 = 302.4 - 216 + 2 = 88.4\ (Ans).$$

(ii) Given, $W_{Ag\ acetate} = 206$ mg, $W_{Ag} = 133.2$ mg, $B = 1$, $M_{HOAC} = ?$
Ans: The working equation is $M_{Ag\ salt} - (B \times 108) + B = M_{acid}$, so,

$$\frac{206\ mg \times (1 \times 108)}{133.2\ mg} - (1 \times 108) + 1 = 167.02 - 108 + 1 = 60.02\ (Ans).$$

(c) Aniline (acidity = 1) $\rightarrow$ Aniline·HCl $\xrightarrow{AgNO_3}$ AgCl $\downarrow$

Calculation: Aniline HCl = W_{salt} = 0.5180 gm
Wt. of AgCl = $W_{Ag\ Cl}$ = 0.5740 gm
How much HCl is present in W_{salt}?

The 0.5740 gm AgCl contains, $\frac{35.5 \times 0.5740\ gm}{143.5} = 0.142$ gm Cl.

Therefore, 0.5180 gm of aniline hydrochloride contains 0.142 gm of Cl since 35.5 gm Cl is present in $\frac{0.5180\ gm \times 35.5}{0.142\ gm} = 129.5$ gm of aniline·HCl, (salt). So the M of aniline is

$$= [M_{aniline.HCl} - HCl] = [129.5 - 36.5] = 93 = M_{aniline}\ (Ans).$$

(c) *The M_{base} is determined by "platinum salt" method:* If the organic base is B, and its acidity is 1, then this will generate B. HCl (Hydrochloride salt of base) which will react with $PtCl_4$ to form $B_2H_2PtCl_6$ (i.e. $2B \cdot HCl + PtCl_4$). Again if the acidity of the base is 2, then the hydrocloride of its base will be $B \cdot (HCl)_2$ which will form platinum salt BH_2PtCl_6 [i.e. $PtCl_4\ B \cdot (HCl)_2$]. These salts on strong heating furnish metallic platinum (at. wt. 195). So at first the $M_{Pt\ salt}$ is determined then M_{base} is calculated. Thus, if,

(i) wt. of $Pt_{salt} = W_{Pt\ salt}$ gm.
(ii) wt. of Pt formed = W_{Ptgm} (at. wt. 195).
(iii) Acidity of the base = 1, 2, etc.
(iv) The mol. wt. of base = M_{base}
(v) The mol. wt. of $Pt_{salt} = M_{Pt\ salt}$.

$$M_{Pt\ salt} = \frac{W_{Pt\ salt} \times 195}{W_{Pt}}$$

$= B_2H_2PtCl_6$ (when B is monoacidic).

Therefore,

$$M_{base} = \frac{1}{2}(B_2H_2PtCl_6 - H_2PtCl_6)$$

Again if the B is dibasic,
the $M_{Pt\ salt} = B \cdot H_2PtCl_6$
Therefore, $M_{base} = (BH_2PtCl_6 - H_2PtCl_6)$.

Examples:

(i) Acidity of the organic base = 1, $W_{Pt\ salt} = 0.652$ gm, $W_{Pt} = 0.215$ gm
What is M_{base}?

Ans: From the working equation $M_{base} = \frac{1}{2}(B_2H_2PtCl_6 - H_2PtCl_6)$, when the acidity of the base is 1 (given).

$$M_{Pt\ salt} = \frac{0.652\ gm \times 195}{0.215\ gm} = 591.3. \text{ Therefore, } M_{base} = \frac{1}{2}(591.3 - 410)$$
$$= 90.65\ (Ans).$$

(ii) The acidity of caffein is 1. The platinum salt of caffein (150 mg) furnishes on strong heating Platinum (36.9 mg). What is the molecular weight of caffein?
Ans: Let caffein is "caf." then its Pt salt will be $Caf_2H_2PtCl_6$ (the acidity of caf = 1, given). Therefore, $M_{Pt\ salt} = \frac{W_{Pt\ salt}}{W_{Pt}} \times 195 = \frac{150\ mg \times 195}{36.9\ mg} =$ $Caf_2H_2PtCl_6$. So $M_{caf} = \frac{1}{2}\left(\frac{150 \times 195}{36.9} - 410\right) = \frac{1}{2}(792.7 - 7.410) = \frac{1}{2} \times$ $382.7 = 191.3$ (*Ans*).

[C] Instrumental methods of molecular weight determination

(a) **Mass Spectroscopy.** (vide 2.12).
(b) **Cryoscopic method** (Depression of freezing point) of compounds by added organic compound).
(c) Ebullioscopic method (Elevation of boiling point of solvent by solute).
(d) Osmotic pressure measurement.

The description of these methods may be had of in good textbooks of physical chemistry. They are rarely used now, hence omitted in this book.

3.14 COMBUSTION OF HYDROCARBONS (EUDIOMETRY)

The mixture of hydrocarbon and oxygen after combustion forms CO_2 and H_2O. If the formula of the hydrocarbon (HC) is C_xHy, then 1 molecule of HC, will generate x CO_2 molecules after combustion. Thus,

$$\underset{[A]}{CxHy} + \underset{[B]}{\left(x + \frac{y}{4}\right) O_2} = \underset{[C]}{x\,CO_2} + \underset{[D]\text{ as steam.}}{\frac{y}{2} H_2O} \quad \text{...eqn. I}$$

The steam [D] formed will condense at room temperature into water. Its volume may be ignored.

If the volume of the HC at NTP is V ml, then the eqn. I will be

$$V \times Cx\,Hy \quad + V\left(x + \frac{y}{4}\right)O_2 = V \times xCO_2 \quad + \quad V.\frac{y}{2}H_2O \qquad \text{...eqn. II}$$

[A] Vol. of HC. [B] Vol. of oxygen consumed [C] Vol. of CO_2 formed [D] Vol. of steam formed

(i) In a closed vessel the known volumes of HC, (V ml) and O_2, $\left[V \times \left(x + \frac{y}{4}\right) ml\right]$ at N.T.P. are taken and after combustion the HC (Vml) and oxygen $\left[V \times \left(x + \frac{y}{4}\right) ml\right]$ will disappear while V ml of CO_2 will be fromed.

The volume of steam $\left(V \times \frac{y}{2}\ ml\right)$ is ignored as after cooling to room temperature (NTP) the steam becomes water of negligible volume. Therefore,

$$V \times CxHy\ (lost) + V\left(x \times \frac{y}{4}\right)O_2\ (lost) = V \times x\ CO_2\ (formed) + V \times \frac{y}{2}\ (formed)$$

(ml) (ml) (ml) *(becomes O) ml*

There is a contraction of volume (called first contraction) which is

$$\left(V + Vx + V\frac{y}{4} - Vx\right) \text{ or } V + \frac{V \times y}{4} \text{ or } V\left(1 + \frac{y}{4}\right) = \textit{first contraction.}$$

(ii) After the combustion the product is treated with alkali to absorb the CO_2 leading to the *second contraction* of V.x ml. at. N.T.P.

(iii) The *vapour density* of the HC if taken as D, then 2D = M (mol. wt. of HC).

(iv) If excess of oxygen is taken (X ml) then the actual volume of oxygen involved in the oxidation can be obtained by subtracting the residual volume (Y ml.) of oxygen from the initial volume of oxygen taken. Thus (X – Y) ml. is consumed $= V\left(x + \frac{y}{4}\right)$ when the formula of HC, is taken as CxHy.

Hence in eqn. II if [A], [B] and [C] are known or [A], [D density] and [C] or [A], [M] and [C] or [A], [X], [Y] and [C] are known then the formula CxHy can be determined by simple arithmetics. In this connection it should be remembered that the former eqn. $\% E = \frac{NO_E \times At.wt._E \times 100}{M}$ may be useful in some occasion.

Examples:

(a)

C_3H_8 (10 ml = V) $(C_xH_y, x = 3, y = 8)$

All volumes at NTP

- Complete Oxidation → Requires $O_2 \left[V\left(x + \frac{y}{4}\right)\right]_{ml}$ $= 10\left(3 + \frac{8}{4}\right)$ ml = 50 ml.
- CO_2 formed → $V \cdot x \cdot \text{ml} = 10 \times ´3 \text{ ml} = 30 \text{ ml.}$
- 1st. Contraction → $V \cdot \left(1 + \frac{y}{4}\right) \text{ml} = 10 \times \left(1 + \frac{8}{4}\right) = 30 \text{ ml}$
- 2nd. Contraction → $V \cdot x \cdot \text{ml } (CO_2) = 10 \times 3 \text{ ml} = 30 \text{ ml.}$
- Vapour denisity D → $\frac{M}{2} = \frac{44}{2} = 22.$
- Percentages of the elements → $\%C = \frac{12 \times 3 \times 100}{44} = 81.2$ and $\%H = \frac{1 \times 8 \times 100}{44} = 18.1.$

(b) 12 ml of HC + excess of $O_2 \xrightarrow{\text{combustion}}$ 1st contraction 30 ml, 2nd contraction 24 ml, what is the formula of HC?

Ans: The general eqn. of combustion is $V \times C_xH_y + V\left(x + \frac{y}{4}\right)O_2 = V \times x$ $CO_2 + V\frac{y}{2}H_2O$. Given, V = 12 ml,

1st contraction 30 ml = $\left[V \cdot \left(1 + \frac{y}{4}\right)\right]$ ml $= 12\left(1 + \frac{y}{4}\right)$ or $12 + \frac{12y}{4} = 30.$

So, y = 6.

2nd contraction 24 ml, $= V \times x\ CO_2$ or $12 \cdot x = 24$, So x = 2.

Hence HC. is C_2H_6 (ethane). *Ans.*

(c) HC taken 20 ml + excess of $O_2 \xrightarrow{\text{combustion}}$ contraction 60 ml, D of HC. = 22. What is the formula for HC?

Ans: The general eqn. of combustion of HC is $V \cdot CxHy + V\left(x + \frac{y}{4}\right)O_2 =$ $V \cdot x\ CO_2 + V \cdot \frac{y}{2}H_2O.$

Given V = 20 ml, 1st contraction $= V\left(1+\frac{y}{4}\right) = 60$ ml, D = 22 (so M = 44).

Therefore, $20\left(1+\frac{y}{4}\right)$ ml = 60 ml or $\frac{20y}{4} = (60-20)$ or 5y = 40. So, y = 8.

The partial formula is C_xH_8. The M = 44. So 12x = 44 − 8 = 36, or x = 3.
Hence, the formula of HC. is C_3H_8.

(d) A mixture of $(CO + CH_4 + H_2)$ = 100 ml taken $\xrightarrow[\text{combustion}]{+O_2 (300\text{ ml})}$ Volume 285 ml. $\xrightarrow{KOH}$ Volume 205 ml (O_2)

What is the volumetric composition of the mixture?

Ans: Let vol. of CO = x ml, CH_4 = y ml and H_2 = z ml.

Given (x + y + z) = 100 ml – eqn. (i)

Residual O_2 = 205 ml. So, O_2 consumed = (300 – 205) ml = 95 ml (ii)

KOH absorbs CO_2 = (285 – 205) ml = 80 ml (iii)

Equations of combustion:

(iv) $2CO + O_2 = 2CO_2$

$x \quad + \frac{x}{2} \quad x$

(v) $CH_4 + 2O_2 = CO_2 + 2H_2O$

$y \quad +2y \quad y$

(vi) $2H_2 + O_2 = 2H_2O$ ... (ml)

$z \quad + \frac{z}{2} \quad 0$

The volumetric composition of the mixture is

CO = 50 ml.
CH_4 = 30 ml. (*Ans*)
H_2 = 20 ml

Calculation:

Therefore (x + y) = 80 ml (eqn. iii) and (iv + v)

Again (x + y + z) = 100 ml (eqn. i)

Therefore, z = (100 – 80) ml = 20 ml = **vol. of H_2**

Again $\frac{x}{2} + 2y + \frac{z}{2} = 95$ ml (eqn. ii) and (iv, v, vi).

or $\frac{x}{2} + 2y + \frac{20}{2} = 95$ ml

or $\frac{x}{2} + 2y = 85$ ml or x + 4y = 170 ml

and x + y = 80 ml

So 3y = 90 ml or **y = 30 ml.** (*Ans.*)

Since x + y = 80 ml (*vol. of CH_4 = y,*)

y = 30 ml

So x = 50 ml = **vol. of CO.** (*Ans*)

3.15 ANALYSIS OF FATS AND OILS

Identification through the following data.

(a) Saponification number: Determined in the following way:

Fat or oil (1 gm) (Starting material) $\xrightarrow{\text{Amount of KOH expressed}}$ Glycerine + K–salt of fatty acids. (Product)

in *milligram* capable of complete hydrolysis of 1 gm of fat or oil. (this quantity is Saponification numbers).

Example: Glycerol tristearate

$$(C_{17}H_{35}-CO_2)_3C_3H_5 + 3KOH \xrightarrow[\Delta]{\text{hydrolysis}} 3C_{17}H_{35}-CO_2K + C_3H_8O_3$$

Mol. wt. = 890 (3 × 56) gm (K–salt of stearic acid) (Glycerol).

Saponification number of glycerol tristearate = $\frac{3 \times 56 \times 100}{890}$ = 189. So this starting material (1 gm) will require 189 mg. of KOH in aqueous solution to completely hydrolyse the fat leaving no residual fat or KOH.

(b) Iodine number: This indicates the unsaturation (C = C) in the substrate.

Fat or oil (100 gm) (substrate) $\xrightarrow{}$ Completely saturated product.
(Amount of iodine expressed in *gm* required to react with (C=C) of the substrate leaving no unsaturation unreacted)

(c) Reichert–Meissl Number (R.M. No.): Amount of water-soluble and volatile acids (generally with approximately C_8 carbon chain) present in *5 gm* of the fat.

Fat (5 gm) substrate $\xrightarrow[\text{2) Steam distillation}]{\text{1) Acidic hydrolysis}}$ The water-soluble steam volatile organic acids are collected.

↓ Neutralised with 0.1 (N) KOH. The number of *ml* of this KOH to neutralise the acid formed above.

K-salt of steam volatile acids.

(d) Table of data of some common oils.

Fat or oil	*Sap. No.*	*I. No.*	*R.M. No.*
(i) Coconut oil	253	(6 –10)	(6 – 7)
(ii) Butter oil	(210 – 230)	(26 – 28)	(17 – 34)
(iii) Peanut oil	(186 – 194)	(88 –98)	0.4

The given fat or oil are analysed and the aforementioned data will indicate the nature (identity) and purity of the sample.

3.16 ESTIMATION OF REDUCING SUGARS (ALDOSES AND KETOSES) WITH FEHLING'S SOLUTION

[A] Reducing sugars are oxidised by Fehling's solution and the end point is colourless solution with the precipitation of red Cu_2O. Cane sugar cannot be determined by Fehling's solution, so cane sugar is hydrolysed into glucose and fructose then estimated by Fehling's solution in the usual way.

[B] Reagents:

(i) Fehling's solution I (orA): A.R. $CuSO_4 \cdot 5H_2O$ (34.60 g) is dissolved in H_2O, a few drops of H_2SO_4 are added to it and the volume is made upto 500 ml.

This solution is stored in a glass stoppered bottle (the stability of the solution is poor).

(ii) Fehling's solution II (or B): Rochelle salt (Sod. Pot. Tartrate, 173 g.) and NaOH (50 g) are dissolved in water and the volume is made upto 500 ml. This solution is stored in a rubber (*not glass*) stoppered bottle.

(iii) Methylene blue indicator (1% aqueous solution) is used to determine the end point.

(iv) Distilled water is always used after boiling the water for 12 hrs. over $KMnO_4$ and redistilled.

[C] Procedure:

(i) A standard solution (approximately 1%) of AR. glucose (preserved in a desiccator) is prepared by accurately weighing and dissolving in a volumetric flask (250 ml) to make up the exact volume with H_2O. The weight of sugar taken = Wgm.

(ii) Exact volume (25 ml) of Fehling's solutions I (F.S.I.) is taken in a 250 ml. clean, dry Erlenmeyer flask using a pipette.

(iii) Almost equal volume (25 ml) of F.S. II is taken in a measuring cylinder and added to the above F.S.I. followed by 50 ml of boiled water. The deep blue copper complex is the reagent. A few boilizers are added to this mixture and a paper collar is fastened on the neck of the Erlenmeyer flask to swirl the boiling reaction mixture. This complex should be prepared just before use.

(iv) The standard glucose solution is added dropwise from a burette to the mildly boiling Fehling's solution mixture (iii) with continuous swirling by holding the conical flask by the collar. The initial deep-blue colour of the reagent gradually fades and the red ppt. of Cu_2O collects at the bottom. Towards the end of the reaction 3–5 drops of methylene blue indicator are added (from a dropper) to the boiling reaction mixture (the boiling should never be discontinued before the end of the reaction). Then 3–4 drops of glucose are added at a time and the reaction mixture is swirled well until the reaction mixture becomes colourless at the end point. The titre-value is noted and the burner removed. This gives the rough idea of the end point.

(v) The above procedure is repeated (at least thrice to get the reliable mean value) by adding the *standard glucose solution* in a continuous stream near the rough titre value (i.e. within 1 ml). Then the sugar is added dropwise (2 drops at a time) to the nearest end point, 3–4 drops of methylene blue are added and the titration is completed to the colourless end point (a piece of white paper may be helpful for this purpose). The exact titre-value is noted and the burner removed. This titre-value is V_k (i.e. volume of the known sugar solution).

[D] Procedure to determine unknown sugar:

(i) The volume of the supplied sample is made upto 250 ml in a volumetric flask with water to make the strength of the solution approximately equal to that of the known glucose solution.

(ii) This sugar solution is taken in a burette. The F.S. mixture is made in a 250-ml conical flask taking F.S.I (25 ml, pipette) and F.S II (25 ml, measuring cylinder) just before use and titrated with the sugar *solution of unknown strength* in the aforementioned procedure (C, iv). Let this volume be V_{un} (i.e. the volume of the unknown sugar (supplied) solution needed to reach the end point).

[E] Calculation: Let 250 ml solution contains Wgm of glucose, then V_k ml will contain $\frac{W_g \times V_k}{250\ ml}$ of glucose. This V_k ml is equivalent to 50 ml of F.S. mixture (25 ml F.S I + 25 ml F.S II) which is also equivalent to V_{un} ml of unknown sugar solution (250 ml). Hence V_{un} contains $\frac{W_g \times V_k\ ml}{250\ ml}$ = x gm of sugar.

Therefore, 250 ml of unknown sugar solution will contain $\frac{W_g \times V_k \times 250\ ml}{250 \times V_{un}\ ml}$ of sugar. This is the quantity of the sugar present in the supplied solution.

NOTES: (i) $CuSO_4$ is poisonous, so pipetteing should be done very carefully. Immediate washing with enough water is recommended if $CuSO_4$ solution enters the mouth. (ii) NaOH is very corrosive, so it is measured in a measuring cylinder (25 ml capacity) instead of a pippette. FS II does not come into the equation of the oxidation. Only copper ion are important factor. (iii) The reaction is a relative one (not absolute), as the unknown sugar is measured against a standard sugar solution by comparison. The reaction is not stoichiometric. (iv) Correct end point determination is very difficult and the handling of the boiling reaction mixture is hazardous. The following method, though expensive, is less hazardous and more easy to detect the end point very accurately (vide 3.19), (v) Fehling's solution I should not be stocked in large quantity as it tends to decompose with time, (vi) 1 ml of this Fehling's solution ≡ 0.00501 g of glucose (approx.).

Example:

(a) Known glucose solution contains 2.1265 g (Wg) of glucose in 250 ml. This solution requires 17.6 ml (V_k ml) to reach end point with 50 ml of F.S. mixture. The unknown titre value (V_{un}) is 23.2 ml, required to reach the end point with 50 ml of F.S. mixture. The unknown sample had been dissolved in water to make the volume 250 ml in a volumetric flask. What is the total quantity of the unknown sugar?

Ans: 250 ml of the known solution contains 2.1265 (Wg) gm of glucose.

∴ 17.6 ml (V_k) of this solution contains $\frac{2.1265\ g \times 17.6\ ml}{250\ ml}$ = x g of glucose.

50 ml of F.S. mixture consumes x gm of glucose.

The same vol. of F.S. mixture (i.e. 50 ml) consumes glucose in 23.2 ml of unknown solution. The weight of the glucose in this 23.2 ml must be *x* gm, i.e., 23.2 ml of unknown sugar solution contains x g of glucose.

$$\therefore \text{ 250 ml of this solution will contain } \frac{x\text{g} \times 250 \text{ ml}}{23.2 \text{ ml}} = \text{y gm of glucose.}$$

$$\text{From the given data, } y_g = \frac{2.1265 \text{ g} \times 17.6 \times 250}{250 \times 23.2} = 1.6132 \text{ g } (\textit{Ans}).$$

(b) Known soln. = 1.9231 gm/250 ml; 50 ml of F.S. mixt ≡ 21.3 ml (V_k) of this soln. and 50 ml of F.S. mixture ≡ 19.7 ml (V_{un}) of unknown solution (250 ml). What is the total quantity of the supplied sugar in the sample?

Ans: $\frac{1.9231 \text{ g} \times 21.3 \text{ ml } (V_k) \times 250 \text{ ml}}{250 \text{ ml} \times 19.7 \text{ ml } (V_{un})}$ = sugar present in g of the supplied sample = 2.079 g. (*Ans*).

3.17 DETERMINATION OF CANE SUGAR (SUCROSE)

[A] It is a non-reducing disaccharide, hence to be estimated after converting it into the reducing sugar which is an equimolar mixture of glucose and fructose. This is done by acid hydrolysis when the optical rotation of cane sugar changes from dextro to levo, so the product is often referred to as invert sugar. Thus,

$$\underset{\substack{\text{(Cane sugar,}\\ \text{mol. wt. 342)}}}{C_{12}H_{22}O_{11}} \xrightarrow[\Delta,\ 30 \text{ min steam bath.}]{[HOH],\ 0.5N\ HCl} \underbrace{\underset{\text{glucose}}{C_6H_{12}O_6} + \underset{\text{fructose}}{C_6H_{12}O_6}}_{\text{Total 360}} \text{ (invert sugar)}$$

[B] Procedure:

(i) About 1 g of sucrose is dissolved in distilled H_2O (20 ml), treated with 0.5 N HCl (10 ml) and heated on a steam bath for 30 mins. The reaction mixture is cooled, neutralised with dil. NaOH and diluted to 250 ml in a volumetric flask with boiled out H_2O.

(ii) The Fehling's solution mixture (25 ml of F.S.I + 25 ml of F.S. II) is standarised with standard glucose solution in the aforementioned method (C, ii).

(iii) In an Erlenmeyer flask (250 ml) the F.S. mixture (50 ml) is taken and completely reduced by the invert sugar solution (Bi) in the usual way. Let the volume be V_{un} ml.

[C] Calculation: Let the standard glucose solution contain W_g of glucose in 250 ml of the solution and each portion of 50 ml Fehling's solution mixture requires V_k ml of this standard solution and V_{un} ml of invert sugar in separate experiments.

Since V_k ml of glucose contains $\frac{W_g \times V_k \text{ ml}}{250 \text{ ml}}$ of glucose (=y g), which is also present in the V_{un} ml of invert sugar.

Therefore, 250 ml of invert sugar (i.e. total cane sugar supplied/taken) is

$$\frac{y \text{ g} \times 250 \text{ ml}}{V_{un} \text{ ml}} = \frac{W_g \times V_k \text{ ml} \times 250 \text{ ml}}{250 \text{ ml} \times V_{un} \text{ ml}} = \frac{W_g \times V_k}{V_{un}} = z \text{ g.}$$

Since 360 g of invert sugar is obtained from 342 g of cane sugar on hydrolysis, z g. of invert sugar is equivalent to $\frac{z \times 342}{360}$ g of cane sugar (= z × 0.95 g).

Example: The standard glucose solution contains 2.1265 g in 250 ml. The volume of invert sugar required to reduce 50 ml of F.S. mixture is 16.7 ml and the standard glucose solution reguires 18.1 ml to reduce the same volume of F.S. mixture. The total volume of the invert sugar solution made is 250 ml. What is the total quantity of the cane sugar supplied?

Ans: 18.1 ml of standard glucose solution contains $\frac{2.1265 \text{ g} \times 18.1 \text{ ml}}{250 \text{ ml}}$ of glucose and 16.7 ml of invert sugar also contains the same amount of reducing sugars, hence, 250 ml of invert sugar will contain $\frac{2.1265 \text{ g} \times 18.1 \text{ ml} \times 250 \text{ ml}}{250 \text{ ml} \times 16.7 \text{ ml}} = 2.3047$ g. Therefore, total cane sugar supplied is $\frac{2.3047 \text{ g} \times 342}{360} = 2.189$ g (*Ans*).

Or $\left(\frac{2.1265(W_g) \times 18.1 \text{ ml } (V_k) \times 250 \text{ ml} \times 342}{250 \text{ ml} \times 16.7 \text{ ml } (V_{un}) \times 360}\right)$ i.e. $\frac{W \times V_k \times 0.95}{V_{un}}$ g of total cane sugar.

NOTE: The concetration of the dil. HCl should be very near to the specified one (i.e. 0.5 N).

3.18 DETERMINATION OF GLUCOSE AND SUCROSE PRESENT IN A MIXTURE BY FEHLING'S SOLUTION

(i) A standard solution of glucose (W_1 g in 250 ml) is titrated against 50 ml of Fehling's solution, using methylene blue as the indicator. Thus the glucose equivalent of the 50 ml of F.S. is known.

(ii) The supplied sample (mixture of glucose and sucrose) is made upto 250 ml with distlled water and titrated against 50 ml of the *same F.S.* Thus the *total amount* of glucose in the supplied sample (which is made upto 250 ml) may be calculated (Wg of glucose).

(iii) From the residual stock solution (in ii) 50 ml is hydrolysed with 25 ml of dilute HCl (0.5 N, should not exceed this strength) by heating on a steam bath for 40 mins. the R.M. is cooled, diluted to 250 ml (volumetric flask) with freshly boiled H_2O and titrated against the *same* F.S. mixture (50 ml). Thus the total reducing sugar (glucose + invert sugar) in original stock solution (ii) is calculated (W_{total} g).

(iv) So the amount of *glucose* in the sample (given) is Wg.
and the amount of *sucrose* in the sample (given) is $(W_{total}\ g - W_g) \times 0.95$.

3.19 ESTIMATION OF GLUCOSE BY VANADATE IN PERCHLORIC ACID

[A] This method is applicable to any water soluble sugar and superior to the Fehling's solution method because it is very easy to perform and gives very accurate results. This is also a relative method and involves the standardisation of the vanadate solution by the same sugar of known strength.

[B] Reagents

(i) Pure glucose is stored over conc. H_2SO_4 in a desiccator for at least one week.

(ii) Distilled water must be boiled for 10–12 hrs. over $KMnO_4$, then resdistilled (D.D. Water).

(iii) Freshly prepared starch solution should be used as the indicator.

(iv) Ammonium (or sodium) vanadate, perchloric acid (70%), potassium iodide, potassium dichromate are all of G.R. grade.

(v) Sodium thiosulphate pentahydrate and hydrochloric acid are of A.R. grade.

(vi) Glucose solution of known strength is prepared by dissolving accurately weighed glucose (0.4–0.7 g) in (double-distilled) water (100 ml).

[C] Procedure:

(i) $Na_2S_2O_3 \cdot 5H_2O$ (25 g) in H_2O (400 ml) is left at room temperature for 48 hrs. Then it is filtered and diluted to 1 L. The solution (about 0.1 N) is standardised in the usual way with standard $K_2Cr_2O_7$ solution.

(ii) Ammonium (or sodium) vanadate (0.5 mole) is dissolved in hot H_2O (200 ml), cooled and very slowly added to the cold solution of perchloric acid (70%. 170 ml dissolved in 200 ml of H_2O) with continuous stirring when pale yellow vanadate solution is obtained. This is stored in a glass stoppered flask. The stability of this solution is excellent.

(iii) This vanadate solution (30 ml, burette) is taken in a standard jointed round-bottomed flask (100 ml) heated under reflux (40 mins.) over a small flame, cooled, treated with KI (2.5 g), left for 5 min. in dark and swirled occasionally. The liberated iodine is titrated with the standardised thiosulphate solution (starch indicator). The same experiment minus the refluxing gives the same result showing that the vanadate is quite stable to the refluxing temperature in perchloric acid solution *when D.D. water is used.*

(iv) In a standard jointed round-bottomed flask (100 ml) fitted with a small condenser is taken the glucose solution (2 ml) of *known strength* from the pipette. The vanadate solution (30 ml, burette) is added to it, the reaction mixture is heated under reflux (40 mins) and the resultant green blue solution cooled to room temperature. The condenser is washed down with D.D. water (10 ml, *pipette*). Then the condenser is removed and the contents of the flask is treated with KI (2.5 g), swirled well, left at room temperature in dark and the liberated iodine is titrated with standard thiosulphate in the usual way (starch indicator). The stable end point is abrupt change of colour from purple to deep sky-blue.

(v) The aforementioned procedure (iv) is adopted for estimation of glucose solution of *unknown strength* (the supplied solution). Best results are obtained when the concentration of the supplied sugar solution is nearly 0.5%. Only two ml. of sugar solution is used for each determination to furnish excellent results.

[D] Calculations:

(i) Wg = weight of glucose in 100 ml of H_2O to make the solution of *known strength*.

(ii) Z ml = Volume of thiosulphate required for blank titration of 30 ml of vanadate solution.

(iii) X ml = Volume of same thiosulphate required for titration of sugar solution of *known strength* (2 ml, pipetted from 100 ml of standard glucose solution).

(iv) Y ml = Volume of same thiossulphate required for titration of sugar solution of *unknown strength* (i.e. supplied sample, 2 ml, pipetted from 100 ml of solution).

$$\text{Total glucose present in unknown sample} = G_g = \frac{W_g(Z-Y)}{(Z-X)}.$$

NOTES:

(i) Cane sugar (sucrose) can be determined with equal accuracy in the aforementioned procedure using standard sucrose solution.

(ii) Very small amount of glucose (≈ 5 mg/ml) can be determined accurately.

(iii) The end point is sharp, easily detectable and stable when the water added during washing is strictly controlled (at 10 ml of H_2O).

(iv) Several experiments can be carried out simultaneously in the non-hazardous method in any laboratory.

3.20 DETERMINATION OF ACETONE

[A] Acetone reacts with known excess of I_2 solution in alkaline solution to furnish CHI_3. The residual I_2 is titrated with 0.1 (N) $Na_2S_2O_3$ using freshly prepared starch solution as the indicator.

[B] Reagents:

(i) Iodine solutions (0.1N): Iodate-free pure KI (20 g) is dissolved in distilled H_2O (40 ml) in a glass stoppered volumetric flask (1 L) and I_2 (AR. 12.7 g) is added to it. The flask is stoppered and shaken well to dissolve the I_2 completely, then the contents are diluted to 1 L with distilled H_2O. If required the solution may be stored in a glass stoppered dark colored bottle for future standardization with standard $Na_2S_2O_3$ solution (0.1 N).

(ii) NaOH solution is prepared by dissolving NaOH (40 g) in distilled H_2O (1 L) with shaking and stored in a rubber stoppered bottle.

(iii) $Na_2S_2O_3$ solution (0.1N) [vide 3.19[C](i)]. The solution is standarised with $K_2Cr_2O_7$ solution in the usual way.

(iv) Freshly prepared starch solution. [vide 3.22[B](iii)]. Two drops are used for each experiment.

[C] Procedure:

(i) A.R. Acetone (10 g approximately, nearly 12 ml) is dissolved in H_2O (distilled, 500 ml). From this solution 25 ml is taken and diluted to 500 ml once again.

(ii) From the resultant solution 25 ml is taken (this will contain approximately 0.02 g of acetone) and dissolved in distilled H_2O (200 ml) in a glass-stoppered Erlenmeyer flask (500 ml).

(iii) Sodium hydroxide solution (approx. N; 25 ml) is added to the acetone solution (C, ii), swirled for 10 mins. And treated with standard iodine solution (B, i) (50 ml, burette). The reaction mixture is shaken well in the stoppered Erlenmeyer flask and left at room temperature for nearly 15 mins, then acidified with dil. H_2SO_4 (1N approx., 26 ml). The residual iodine is immediately tirated with standard $Na_2S_2O_3$ solution (0.1N) using starch as the indicator.

(iv) The blank experiment is to be done always.

(v) The quantity of acetone in the solution: 1 ml of 0.1N $I_2 \equiv 0.0009680$ gm acetone.

(vi) Equation: $CH_3-CO-CH_3 + 3\,I_2 + 4\,NaOH = CH_3-CO_2Na + 3NaI + 3H_2O + CHI_3$.

3.21 DETERMINATION OF PHENOLS AND ANILINE

[A]

(i) The following iodometric method is adopted to determine the purity of the phenolic compounds (or aniline) in a sample.

(ii) The same procedure (also reagents and calculation) is adopted replacing phenol by aniline after converting the latter into aniline hydrochloride (0.25 g + few drops of HCl).

[B] Reagents:

(i) Bromate-bromide mixture: $KBrO_3$ (A.R., 5.567 g) and KBr (pure, 75 g) are dissolved in distilled water and the volume is made upto 1 L (volumetric flask). The strength is (0.2 N).

(ii) Sodium thiosulphate solution (0.1 N): $Na_2S_2O_3 \cdot 5H_2O$ (25 g) in 1 L of freshly boiled water [vide 3.19[C](i)].

(iii) Pure KI solution (20%): KI (100 g) in distilled H_2O (400 ml).

(iv) Freshly prepared starch solution.

[C] Procedure:

(i) The supplied phenol (0.25 g) dissolved in NaOH solution (5 ml, 10%) is taken and the volume is made upto 250 ml (volumetric flask) with distilled H_2O. Likewise, aniline, (0.25 g with a few drops of conc. HCl in H_2O) may be taken.

(ii) From the above solution (i) 25 ml is taken (burette) in a 500 ml standard jointed (B-24) Erlenmeyer flask.

(iii) The bromate-bromide mixture (25 ml, pipette) is added to it (ii) followed by distilled H_2O (25 ml).

(iv) Conc. HCl (A.R.. 5 ml) is added to (iii) and the flask is stoppered, shaken well and left at room temperature for 30 mins. with occasional swirling.

(v) Carefully (prevent loss of Br_2) KI solution (10 ml, 20%) is added to it (iv).

(vi) The flask is swirled well and allowed to stand at room temperature for 10 mins. in darkness.

(vii) The stopper and inside of the flask is washed down with a little H_2O, and the I_2 liberated by the excess of Br_2 (left after reaction with the phenol (or aniline) is estimated by $Na_2S_2O_3$ (standard, 0.1N) solution using starch as the indicator in the usual way.

(viii) Each time blank experiment is to be carried out in the same way.

(ix) *Equations:*

(a) $KBrO_3 + 5\,KBr + 6\,HCl = 6\,KCl + 3H_2O + 3\,Br_2$

(b) $C_6H_5OH + 3Br_2 = C_6H_2Br_3(OH). + 3HBr$

(c) $C_6H_5NH_2 + 3Br_2 = C_6H_2Br_3(NH_2) + 3HBr.$

[D] Calculations:

Let, V_1 = Vol. of $Na_2S_2O_3$ (blank).

V_2 = Vol. of $Na_2S_2O_3$ (after expt).

N = Normality of thiosulphate

M = Molecular wt. of phenol (or aniline).

W_g = Weight of the sample in gram.

Z = Number of Br atom substituted in the phenol (or aniline).

$$\text{Purity (\%)} = \frac{(V_1 - V_2)\cdot N.M}{W_g \cdot Z \cdot 20}$$

3.22 DETERMINATION OF THE EQUIVALENT WEIGHT OF A CARBOXYLIC ACID BY IODOMETRIC METHOD

[A] The carboxylic acid liberates quantitatively iodine from the mixture of KIO_3 and KI in aqueous solution. This iodine is estimated by standard thiosulphate solution and the equivalent weight of the acid is calculated. The method is expensive but gives very accurate results if carefully performed.

[B] Reagents:

(i) Potassium iodate-potassium iodide mixture:– Prepared by dissolving $K1O_3$ (A.R., 3.0 g) and KI (A.R., 25 g) in distilled water (200 ml).

(ii) Sodium thiosulphate (0.1N) solution prepared from $Na_2S_2O_3 \cdot 5H_2O$ (A.R., 25 g) is standardised in the usual way.

(iii) Starch indicator: A paste of soluble starch (1 g) in water is poured onto boiling water (100 ml), stirred for one min., cooled, KI (3g) is added to it and the resultant indicator is stored in a stoppered bottle. Freshly prepared indicator should always be used.

[C] Procedure:

(i) The pure dry organic acid is accurately weighed (nearly 0.25 g) in an Erlenmeyer flask (250 ml) with a glass stopper and dissolved in sufficient distilled water or 50% pure ethanol if the acid is sparingly soluble in H_2O.

(ii) To the above solution is added the iodate-iodide mixture (20 ml) the flask is stoppered quickly, stirred well and left at room temperature for 10–15 mins in darkness.

(iii) The liberated iodine is immediately estimated by 0.1 (N) $Na_2S_2O_3$ solution using starch solution as the indicator.

(iv) The equation is as follows:

(a) $KIO_3 + 5KI + 6RCO_2H = HIO_3 + 5\ HI + 6R\text{–}CO_2K$.

(b) $HIO_3 + 5HI = 3I_2 + 3H_2O$.

(c) $3I_2 + 6Na_2S_2O_3 = 3Na_2S_4O_6 + 6NaI$.

[D] Calculation: 1 mole of carboxylic acid ≡ 127 × ng of iodine (n = basicity of the acid).

3.23 DETERMINATION OF EQUIVALENT WEIGHT (E) AND MOLECULAR WEIGHT (M) OF CARBOXYLIC ACIDS BY SODIUM HYDROXIDE SOLUTION

[A] Reagents:

(i) A solution of NaOH (0.4 g in freshly boiled H_2O, 500 ml; approximately 0.02 N).

(ii) Phenolphthalein indicator.

(iii) Any one of A.R. succinic acid (E, 59), phenylacetic acid (E,136), benzoic acid (E,122).

[B] Procedure:

(i) In an Erlenmeyer flask (100 ml) accurately weighed (W_1 g) of any pure acid (30 – 35 mg) is taken and titrated with NaOH solution (the end point is very faint pink colour with phenolphthalein indicator). *Let the volume be V_1 ml.*

(ii) Similarly, the pure unknown sample is accurately weighed (W_2 g, 35–40 mg) and titrated with the same NaOH solution. Let this volume be V_2 ml.

(iii) The equivalent weights are E_1 (of known acid) and E_2 (of unknown acid) and M_2 is the molecular weight of the unknown acid.

[C] Calculation:

$$E_2 = \frac{E_1 \times W_2 \times V_1}{W_1 \times V_2}\text{; } M_2 = E_2 \times n \text{ (n = basicity of the unknown acid).}$$

3.24 SEMIMICRO DETERMINATION OF METHOXYL GROUP (ZEISEL METHOD, ALSO CALLED ZEISEL-PRELOG-VIEBOCK-BRECHER METHOD): VOLUMETRIC ANALYSIS

[A] The method involves the quantitative conversion of methoxyl group, (OMe), into methyl iodide (MeI) which generates iodic acid (HIO_3). The latter acts with hydroiodic acid (generated from KI) to furnish iodine quantitatively which is estimated by standard sodium thiosulphate solution in the usual way. The following flow sheet is illustrative of the principle of analysis.

$$\text{Accurately weighed R-OMe (methoxy compound) (non-volatile)} \xrightarrow[\text{boiling)}]{\text{HI (constant}} CH_3I \xrightarrow[+\ NaOAC]{Br_2/HOAC} \text{(i)} \quad \text{(chased by } CO_2\text{)}$$

$$\text{(i) } IBr \xrightarrow{Br_2} \underset{\text{(Iodic acid)}}{HIO_3} \xrightarrow{HI} I_2 \text{ (estimated by standard } Na_2S_2O_3 \text{ soln.)}$$

(B) Equations:

(i) $R–O–CH_3 + HI = R–OH + CH_3I$ [Reagents (i), (iii), (xvii)] (R>>Me)

(ii) $CH_3I + Br_2 = CH_3Br + IBr$ (iodine monobromide) [Reagent (v)]

(iii) $IBr + 2Br_2 + 3H_2O = HIO_3$ (iodic acid) + 5 HBr [Reagent (v)]

(iv) $HIO_3 + 5HI = 3H_2O + 3I_2$ (iodine), [Reagents (xx), (xv)].

(v) $3I_2 + 6Na_2S_2O_3 = 6NaI + 3\ Na_2S_4O_6$ (sod. tetrathionate), [Reagents (ix), (xi)].

(vi) $Br_2 + HCO_2H = 2HBr + CO_2$ (excess of Br_2 decomposed in HIO_3)

(C) Calculations: $OMe \equiv MeI \equiv HIO_3 \equiv 3I_2 \equiv 6Na_2S_2O_3$.

And,

$$\% \text{ (OMe)} = \frac{V \times N \times 517.16}{W \text{ (mg)}}$$

or 1 ml of 1 (N) thiosulphate $\equiv$ 0.005171 gm of (OMe)

V = vol. (ml) of standard thiosulphate required.

N = Normality of the thiosulphate solution.

W_{mg} = Weight of the sample (in mg) analysed.

[D] Reagents: (All regents are A.R. and solvents are pure).

(i) Constant boiling hydrodic acid = (57% HI + 43% H_2O), b.p. 127°C/760 mm/Hg, Sp gr. 1.7.

(ii) Elemental bromine (b.p. 58.78; density 3.12 at 20°C).

(iii) Phenol (b.p. 182°C) and propionic anhydride (bp.116°C).

(iv) Glacial acetic acid (freshly distilled, b.p. 118°C).

(v) NaOAc + HOAc buffer (10%) solutions with Br_2: Made from anhydrous NaOAc (10g) + gl. HOAc (95 ml) + H_2O (5 ml). Added 0.3 ml Br_2 per 10 ml of buffer soln.

(vi) Red phosphorus (preserved under H_2O).

(vii) Sodium acetate (fused and preserved before use).

(viii) Formic acid (b.p. 100.5°C).

(ix) Sodium thiosulphate solution (0.05 N, order) from $Na_2S_2O_3 \cdot 5H_2O$ [vide 3.19 [C](i)].

(x) Potassium dichromate *standard solution* (0.05 N, order) from $K_2Cr_2O_7$.

(xi) Starch indicator (freshly prepared, vide 3.22 [B, iii]).

(xii) Kipp's apparatus (checked for a steady supply of CO_2 gas purified by passing through $NaHCO_3$ washer).

(xiii) A 8–10% aqueous solution of antimonyl sodium tartrate + suspension of red phosphorus in water *or* red phosphorus in H_2O or cadmium sulphate with red phosphorus in water (used in wash tower to arrest HI, I_2 and H_2S chased by CO_2).

(xiv) A 25% NaOAc aqueous solution.

(xv) A 10% (v/v) aqueous solution of sulphuric acid.

(xvi) Silicone stop-cock grease.

(xvii) Redistilled propionic anhydride (b.p. 169°; density 1.03).

(xviii) Carborundum chips or other boiling chips.

(xix) Thin aluminium foil for making small weighing cups.

(xx) Potassium iodide.

[E] Plan of apparatus fitting:

(Kipp's Appartus) $CO_2\uparrow$ ⟶ Washer (Aq. $NaHCO_3$ inlet tube just dipped; 0.5 cm). ⟶ Zeisel flask (sample + HI + solvents) ⟶ Liebig condenser (HI↓; MeI↑)

↓

Hard glass test tube (hanging by a thread) containing reagent (V). [$HIO_3\downarrow$, $Br_2\downarrow$ ⟵ Delivery tube with spiral (MeI↑, $CO_2\uparrow$) ⟵ Washer (contains reagents xiii) HI↓, $I_2\downarrow$; MeI↑, $CO_2\uparrow$).

↓

Erlenmeyer flask [solution processed and idometric estimation by standard sodium thiosulphate solution (0.05N, order)]

[F] Procedure: All reagents are taken from [D].

(i) In the cup made from aluminium foil is taken the accurately weighed (ca 20 mg) sample and inserted carefully into the Zeisel flask.

(ii) Phenol (0.5 g), propionic anhydride (1 ml) and boiling chips are added in the Zeisel flask (i) to dissolve the sample by warming, if necessary.

(iii) To the cold reaction mixture (ii) is added HI (ca 10 ml), quickly connected to the outlet tube of Kipp's apparatus (CO_2 gas washed *via* aqueous $NaHCO_3$ tower) and the condenser is fixed on the top of the Zeisel flask (water flow started) followed by connecting the washer containing the suspension of red phosphorus (vide reagents, xiii). Silicone grease or a drop of pure H_2O should be used on each ground-glass joint to make the system leak-proof.

(iv) The delivery tube holding the glass spiral is connected to the outlet of the wash tower containing red phosphorus. Good springs are used on the ears of the standard joints so that the connection did not turn loose.

(v) The hard glass test tube containing the sodium acetate-acetic acid buffer with the dissolved bromine (15 ml, vide Reagent v) is used as the receiver of purified methyl iodide. This test tube is hung from a suitable stand by a strong thread to make its movements free and easy.

(vi) The delivery tube with the spiral (iv) is dipped into the reaction mixture (v), the issuing gas bubbles (ca one bubble per second) at the bottom of the test tube. The gas bubbles take a longer journey around the spiral thus complete absorption of methyl iodide in the buffer solution is ensured. The CO_2 gas escapes the test tube.

(vii) The Zeisel flask is slowly lowered into the oil bath which is heated by a small flame (or a microburner) at 145–150°C for an hour under the slow stream of CO_2 chasing the CH_3I formed towards the absorption tube (V). Then the hard-glass test tube is lowered, the spiral delivery tube is detached and washed inside and outside with pure water, the washings being collected in a glass stoppered Erlenmeyer flask (250 ml) and the contents of the absorption tube is quantitatively transferred into the same Erlenmeyer flask. An aqueous solution of sodium acetate (25%, 10 ml) is added to the contents of the Erlenmeyer flask and the volume is made upto 150 ml with pure water.

(viii) Formic acid is added dropwise (avoiding excess) to the reaction mixture to destroy excess of Br_2. A drop of methyl red indicator may be used to ascertain the complete removal of Br_2. The HBr gas with bromine vapour (if present) should be blown out by cold air.

(ix) To the aforementioned solution (viii) was added KI (1 g) followed by 10 ml of dilute H_2SO_4 (Reagent xv). The flask is stoppered, left for 5–10 mins in darkness and the liberated iodine is titrated with standardised $Na_2S_2O_3$ (ix; by $K_2Cr_2O_7$ standard solution) using starch (xi) as the indicator.

(x) A blank experiment may be conducted using all the reagents – the sample exactly in the same way.

3.25 DETERMINATION OF METHOXYL AND ETHOXYL IN THE SAME SAMPLE (MAKENS, LOTHRINGER AND DONIA MICRO METHOD)

[A] In the Zeisel method HI decomposes—OMe into MeI which is estimated iodometrically *via* HIO_3 (vide 3.24). The same procedure would generate MeI and Et1 from –OMe and – OEt groups, respectively, when present in the same sample. But these iodo compounds cannot be determined unless and until they are separated quantitatively. The fractional distillation of MeI (b.p. 42.5°C) and Et1 (b.p. 72.2°C) would not be suitable in Zeisel method since these alkyl iodides are driven out by a stream of CO_2, so there is a less likelihood of quantitative separation in the vertical Liebig condenser. This problem of separation of MeI and Et1 has been overcome by the use of trimethylamine in anhydrous and resdistilled nitrobenzene as the absorption solvent contained in successive two hard glass test tubes (the second one is used just as a precaution, since it has been observed that no determinable iodide escapes the first absorption tube).

The MeI generates highly insoluble crystalline tetramethyl ammonium iodide ($Me_4N^{\oplus} I^{\ominus}$) in nitrobenzene but the Et1 furnishes very soluble ethyltrimethyl ammonium iodide ($EtMe_3N^{\oplus} I^{\ominus}$) in nitrobenzene solvent. The precipitated $Me_4N^{\oplus} I^{\ominus}$ is separated quantitatively by filtration, the filtrate contains $EtMe_3N^{\oplus} I^{\ominus}$. Both of these ammonium salts are determined either by Volhard method (modified) or by iodometric method (vide 3.24), the nitrobenzene posing no problem. Absolute ethanol may replace nitrobenzene but here the complete precipitation of $Me_4N^{\oplus}I^{\ominus}$ requires almost 24 hours at room temperature. The following flowsheet illustrates the outline of the present procedure.

Accurately weighed sample (10 mg ± 5 mg) [R(OMe)(OEt)] or [R′(OMe) + R″(OEt)] $\xrightarrow[\text{CO}_2\text{, Chasing [3.24]}]{\text{HI}}$ MeI/EtI (mixture) $\xrightarrow[+\,CdSO_4\text{(5\% Soln)}\ [-I_2,\,-HI,\,-H_2S]]{\text{Red } P/H_2O}$ Delivered into the mixture of Me_3N in nitrobenzene

$\longrightarrow$ Crystalline $Me_4\overset{\oplus}{N}\overset{\ominus}{I}$ and soluble $EtMe_3\,\overset{\oplus}{N}\,\overset{\ominus}{I}$ are separated by filtration and determined

[B] Reagents: (All reagents are A.R. and solvents are pure.)

(i) HI (vide 3.24) propionic anhydride and phenol.

(ii) Trimethylamine (0.5 N in nitrobenzene, anhydrous and redistilled).

(iii) Nitrobenzene, dried and freshly distilled.

(iv) Red phosphorus, D.D. water, and 5% solution of $CdSO_4$ in H_2O.

(a) For determination by modified Volhard method:

(v) Nitric acid solution (1:1, v/v in H_2O).

(vi) Silver nitrate solution (standard, 0.01N order, Eq. wt. 169.89).

(vii) Ferric nitrate indicator, saturated solution in dilute HNO_3 (v).

(viii) Potassium sulphocyanide solution (standardised, 0.01N order).

(b) For iodometric titration: [3.24]

(ix) Glacial acetic acid.

(x) Solid potassium acetate.

(xi) Elemental Br_2.

(xii) Aqueous sodium acetate solution (25%).

(xiii) Formic acid (85%)

(xiv) Potassium iodide (10% aqueous solution).

(xv) Sulphuric acid (10% solution).

(xvi) Sodium thiosulphate solution (0.05 N order, standardised)

(xvii) Standard potassium dichromate.

(xviii) Freshly prepared starch solution.

[C] Procedure: (vide 3.24; Zeisel method):

(i) The sample is accurately weighed (ca 10 mg) on a tin foil and taken in the Zeisal flask. The vertical condenser is fixed on this reaction flask, the washer containing the suspension of red P with $CdSO_4$ solution (5%) is placed on the condenser and thus purified alkyl halides are passed (driven by a stream of CO_2, 1 bubble per second) through the delivery tube (surrounded by glass spiral coil) diped into a solution of trimethylamine (0.5N; 4 ml) in nitrobenzene in the first absorption tube. The same absorption solution (3 ml) is taken in the second tube which is connected with the first one.

(ii) The mixture of propionic anhydride and phenol (3–4 drops) is added to the sample in the reaction flask followed by constant boiling HI (4 ml) and a few boiling chips.

(iii) From Kipp's apparatus CO_2 gas is bubbled through the aqueous solution of $NaHCO_3$ (water tower) and slowly passed in the reaction flask.

(iv) The reaction flask is immersed in the oil bath, water is passed through the condenser and temperature of the reaction mixture is raised to 140–145°C and left for 2 hrs. (EtI is taking longer time). The issuing volatile products are washed in red P suspension containing $CdSO_4$ solution and passed into the absorption tube when gradually $Me_4N^{\oplus} I^{\ominus}$ crystallised out (in the first tube) while $EtMe_3N^{\oplus} I^{\ominus}$ remained in the solution.

(v) The absorption tubes are disconnected, left at room temperature for 45 mins. The crytallised product is filtered under suction (water aspirator). The second tube is also treated in the same way and the contents are estimated separately. The solid $Me_4N^{\oplus}I^{\ominus}$ is washed with absloute EtOH (0.5 ml × 2) and dried under suction.

(vi) This solid (v) is dissolved in warm H_2O (5 ml), the containers and funnel are washed down with warm H_2O (1 ml × 5) and the combined solution (ca 10 ml) is analysed in either of the following two ways.

(a) *Determination of $Me_4N^{\oplus} I^{\ominus}$ by modified Volhard method:*

(i) To the above aqueous solution (vi) is added accurately measured $AgNO_3$ solution (0.01N, 5–10ml, from burette) followed by acidification with HNO_3 (1:1, v/v; H_2O) then titrated with standardised potassium thiosyanate solution (0.01N, oder) using ferric nitrate as the indicator (stable end point, faint brown colour).

(ii) In the nitrobenzene solution (first tube) $EtMe_3N^{\oplus}I^{\ominus}$ remained which is successively treated with H_2O (5 ml), standard $AgNO_3$ solution (accurately measured, upto 10 ml, 0.01N, order), dilute HNO_3 (acidification), then titrated with standardised KSCN solution (0.01N, order) with $Fe(NO_3)_3$ as the indicator (stable faint brown colour end point).

(b) *Determination of alkylammonium iodides by iodometric method (vide 3.24).*

(i) To the solid $Me_4N^{\oplus} I^{\ominus}$ obtained after filtration are added successively H_2O (2 ml) glacial HOAC saturated with KOAc (10 ml solution), Br_2 (12 drops), and shaken (5 mins). To this mixture is added aqueous solution of NaOAc (5 ml, 25% solution) and the excess of Br_2 is decomposed by HCO_2H (12 drops, 85% solution) then the reaction mixture is left at room temperature for 3 mins.

(ii) To the above reaction mixture is added a solution of KI (5 ml, 10% solution) followed by dilute H_2SO_4 (5 ml, 10% solution) and the liberated. I_2 is titrated with standardised $Na_2S_2O_3$ solution (0.05 N order) using freshly prepared starch as the indicator in the usual way.

(iii) Similarly the $EtMe_3N^{\oplus} I^{\ominus}$ dissolved in the nitrobenzene obtained after separation of solid $Me_4N^{\oplus} I^{\ominus}$ is estimated by the method described above (steps b, i and b, ii).

[D] Calculations:

(a) 1 ml of 1(N) $AgNO_3$ ≡ 0.1269 gm of I ≡ 0.031005 gm (OMe group) ≡ 0.045004 gm (OEt group).

(b) 1 ml of 1(N) $Na_2S_2O_3$ ≡ 0.005171 g (OMe) ≡ 0.007506 g (OEt).

[E] Notes:

(i) When only OMe or OEt group is present, the filtration step is omitted (step v).

(ii) The sample may be a mixture of R′–OMe and R″–OEt.

(iii) The solubility of $Me_4N^{\oplus} I^{\ominus}$ may be further lowered by adding dry n-heptane

(iv) The percentage error generally lies within ±0.5 (by subtraction).

3.26 SOME INFORMATION AND CALCULATIONS IN ORGANIC CHEMISTRY

[A] The oxidation number of carbons: Calculation is done in the following way:

(i) Elemental carbon has oxidation number zero.

(ii) Carbon connected with another carbon is given zero oxidation number, each hydrogen connected with C, and each heteroatom (e.g. Hal, O, N, S etc.) connected with C are given oxidation number (–1) and (+1), respectively. Double bonded heteroatoms are given (+2) and triple bonded heteroatoms are given (+3) oxidation numbers.

Examples: Compound (oxidation number of the underlined carbons).

CH_4 (– 4), R–CH_3 (– 3), R–CH_2–R (– 2), CH_3–OH (– 3 + 1 = –2), R_3CH (–1),

R_4C(0), CH_2O (– 2 + 2 = 0, ∵ $H_2C=O$, double bonded oxygen), CH_2X_2

(– 2 + 2 = 0), R – CH = 0 (+ 1), R_3C –OH (+ 1), H—C(=O)–OH (–1 + 3 = +2),

$R_2C=O$(+2), R – CO_3H (+ 3), R – C ≡ N(+3), CX_4(+4) CO_2(+4), This shows that the oxidation number of carbon has the range between –4 and +4.

(iii) The total oxidation numbers of each carbon of the molecule used as starting material is counted. The total oxidation numbers of the product is also counted. When these two totals become equal then the reaction undergoing the starting material is neither oxidation nor reduction. The example of such neutral reaction is hydration of olefins. Thus,

CH_3—CH = CH—CH=O $\underset{-\text{HOH}}{\overset{+\text{HOH}}{\rightleftharpoons}}$ CH_3—CH(OH) — CH(H)— CH=O

Ox. No (–3) + (–1) + (–1) + (–1 + 2)

Total for the molecule = (–4)

(–3) + (0) + (–2) + (+1) = Oxidation No. of each carbon.

Total for the mol. = (– 4)

So this reaction is neither oxidation nor reduction.

(iv) If the oxidation number changes from (–) to (+) [decrease of (–) or increase of (+)] the reaction is *oxidation* of the substrate.

Examples

(a) $CH_3-CH_2-OH \xrightarrow{H_2CrO_4/H^{\oplus}} CH_3-C(=O)-OH$

$(-3)+(-1)=(-4) \qquad (-3)+(+3)=0$

(The total Oxidation number of molecules changes from (–4) to 0. Hence the reaction of oxidation

(b) $CH_3-CH=CH-CH_2-CH_3 \xrightarrow{\text{Ozonolysis}}$

$(-3)+(-1)+(-1)+(-2)\ (-3)=(-10)$

The oxidation number of the molecule is (–10)

$CH_3-CH=O+CH_3-CH_2-CH=O$

$(-3)+(+1) \quad + \quad (-3)+(-2)+(+1)=(-4)$

From (–10) to (–2) and (–4). So the reaction is oxidation [the (–) decreases].

(c) $O=CH-CH=O \xrightarrow[\text{ii) } H_3O^{\oplus}]{\text{i) } KMnO_4/HO^{\ominus}}$

$(+1)+(+1)=(+2)$

$HO-C(=O)-C(=O)-OH$

$(+3)+(+3)=(+6)$

The oxidation number increases from (+2) to (+6). Hence the reaction is oxidation.

(v) If the oxidation number decreases [(+) to (–) or decrease of (+) or increase of (–)], the reaction will be *reduction* of the substrate. Thus,

(a) $CH_3-CH=O \xrightarrow{LiAlH_4} CH_3-CH_2-OH$

$(-3)+(+1)=(-2) \qquad (-3)+(-1)=(-4)$

From (–2) to (–4). Increase of (–), so the reaction is reduction of substrate.

(b) $CH_3 - CH = O \xrightarrow[\text{conc. HCl}]{\text{Zn/Hg}} CH_3 - CH_3$ From (–2) to (– 6), so the reaction is reduction.

Total (–2)   (–3) + (–3) = (–6)

(c) $O = CH - CH_2 - COOH \xrightarrow{LiAlH_4}$

(+1) + (–2) + (+3)

Total = (+2)

$HO - CH_2 - CH_2 - CH_2 - OH$, so the reaction is reduction. [∵ (+2) to (–4)].

(–1) + (–2) (–1) = (–4)

[B] Formal charge: The formal charge (F.C.) on any atom of a molecule is calculated as follows:

F.C. = Number of electrons on the outer orbit of the atom minus (the lone pair of electron of this atom + half of the total electrons engaged in bond making). This is explained in the following examples.

(a) Number of electrons in the outermost orbits of the elements in parentheses. H(1), Li(1), Be(2), B(3), C(4), N(5), O(6), Cl(7), S(6), P(5), Si(4), etc.

(b) F.C. = Formal charge on *each* atom of nitric acid (HNO_3), and other molecules and radicals.

Structural formula	Atoms	No. of electrons on the outermost orbit [A]	Lone pair of electrons [B]	Half of bonded electrons [C]	F.C. = [A] – [B + C]
1. 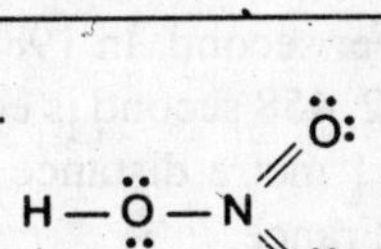 = HNO_3	H·	1	0	$\frac{1}{2} \times 2 = 1$	F.C. (H) = 1 –(0 + 1) = 0
	N	5	0	$\frac{1}{2} \times 10 = 5$	F.C. (N) = 5 – (0 + 5) = 0
	–Ö–	6	4	$\frac{1}{2} \times 4 = 2$	F.C. (O) = 6 – (4 + 2) = 0
	=Ö:	6	4	$\frac{1}{2} \times 4 = 2$	F.C. (O) 6 – (4 + 2) = 0 Net charge = 0

2. NO_3^- (nitrate ion) 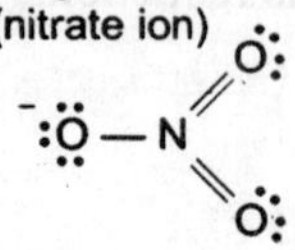	:N:	5	0	$\frac{1}{2}\times 10 = 5$	F.C. (N) = 5 – (0 + 5) = 0
	:O⊖	6	6	$\frac{1}{2}\times 2 = 1$	F.C. (O) = 6 – (6 + 1) = –1
	:O=	6	4	$\frac{1}{2}\times 4 = 2$	F.C. (O) = 6 – (4 + 2) = 0 Net charge = –1
3. 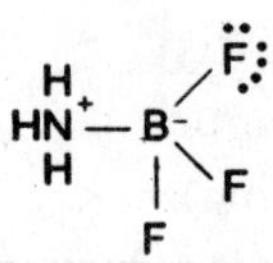	H F N	1 7 5	0 6 0	1 1 4	F.C. (H) = 0 F.C. (F) = 7 – (6 + 1) = 0 F.C. (N) = 5 – (0 + 4) = +1
	B	3	0	$\frac{1}{2}\times 8 = 4$	F.C. (B) = 3 – (0 + 4) = –1 Net charge 0
4. 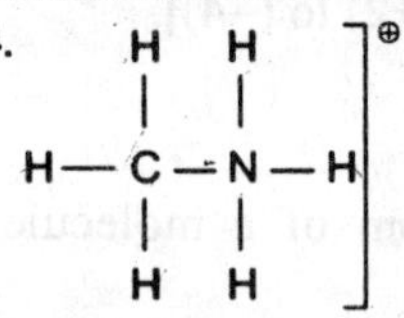	H C N	1 4 5	0 0 0	1 4 4	F.C. (H) = 0 F.C. (C) = 0 F.C. (N) = 5 – (0 + 4) = +1 Net charge = +1
5. $SO_4^{2\ominus}$ 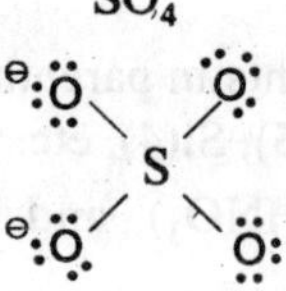	S	6	0	4	F.C. (S) = 6 – (0 + 4) = +2
	O	6	6	$\frac{1}{2}\times 2 = 1$	F.C. (O) = 6 – (6 + 1) = –1 For 4 oxygen atom F.C. = –4 So net change = +2 – 4 = –2 Therefore, the sulphate ion has two negative charges.

3.27 DATA BANK USEFUL FOR THE ORGANIC CHEMISTS

[A] General measurements:

(a) Length: Light travels at the rate of 299, 792, 458 metres per second. In 1983, it had been proposed that the distance covered by light in 1/299, 792, 458 second is equal to 1 metre. Previously a platinum–iridium rod was marked for 1 metre distance and kept at International Bureau of Weights and Measures in Paris, France.

(b) Abbreviations of different measurements: Names (abbreviations):

1. Gram (g),
2. Litre (L),
3. Metre (m), (also meter)
4. Second (s),
5. Centigram (cg),
6. Centilitre (cL),
7. Square centimetre (cm^2),
8. Cubic centimetre (cc^3),
9. Metric carat (cM),
10. Decilitre (dL).

11. Decimetre (dm).
12. Square decimetre (dm^2).
13. Cubic decimetre (dm^3).
14. Square kilometre (km^2).
15. Kg (kilogram).
16. Kilo Joule (kJ).
17. Kilolitre (kL).
18. Kilometre per hour (km/h).
19. Kilosecond (ks).
20. Hour (h).
21. Hectare (ha).
22. Hectolitre (hL).
23. Joule (J).
24. Square metre (m^2).
25. Cubic metre (m^3).
26. Milligram (mg).
27. Megagram (Mg).
28. Milli Joule (mJ).
29. Millilitre (mL).
30. Millimetre (mm).
31. Cubic millimetre (mm^3).
32. Megalitre (ML).
33. Millisecond (ms).
34. Minute (min.).
35. Microgram (μg).
36. Microlitre (μL).
37. Micrometre (μm).
38. Microsecond (μs).
39. Nanogram (ng).
40. Nanometre (nm).
41. Picometre (pm).
42. deci = (d).
43. centi = (c).
44. milli = (m).
45. micro = (μ).
46. nano = (n).
47. pico = (p).
48. femto = (f).
49. atto = (a).

(c) Metric system in Science (litre alternate spelling liter):

(i) Angstrom (**eight** letters in the word) = 10^{-8} cm = 10^{-10} m = Å.

(ii) Micron (**six** letters in the word) = 10^{-6} = micro (μ, mu).

(iii) Milli z = 10^{-3} z (z = litre, metre or second) e.g. millilitre = 10^{-3} litre (z = litre)

(iv) Micro z = 10^{-6} z (z = litre, metre = meter or second)

e.g. microsecond = 10^{-6} second (z = sec).

(v) Nano z = 10^{-9} z (– do –) e.g. nanometre = 10^{-9} metre (z = metre).

(vi) Pico z = 10^{-12} z (– do –) e.g. picosecond = 10^{-12} second.

(vii) Femto z = 10^{-15} z (– do –) e.g. femtosecond = 10^{-15} second

(viii) Atto z = 10^{-18} z (– do –) e.g. attosecond = 10^{-18} second.

(ix) Deca z = 10 z (– do –) = da e.g. decametre = 10 metres.

(x) Hecto z = 10^2 z (– do –) = h e.g. hectometre = 100 metres.

(xi) Kilo z = 10^3 z (– do –) = k e.g. kilogram = 10^3 gm.

(xii) Myria z = 10^4 z (– do –) = my

(xiii) Mega z = 10^6 z (– do –) = M e.g. megawatt = 10^6 Watt.

(xiv) Giga = 10^9 z (– do –) = G

(xv) Tera = 10^{12} z (– do –) = T
(xvi) Peta = 10^{15} z (– do –) = P
(xvii) Exa = 10^{18} z (– do –) = E
(xviii) Zetta = 10^{21} z (– do –) = Z
(xix) Yotta = 10^{24} z (– do –) = Y

NOTES:

(i) **Only one symbol will be used, e.g. in place of millimillimilligram, the word nanogram is used. In place of kilokilogram the word megagram will be used.**

(ii) **Symbol will be always singular, e.g. 1 g, 10 g (not 10 gs), 100 km (not 100 kms), but 100 grams (gram is word not symbol, so plural). So 100 grams = 100 g.**

(d) Length: (Metre); 10^3 m (1 km); 1852 m (1 International nautical mile). 10^{-1} m (1 dm); 10^{-2} m (1 cm); 10^{-3} m (1 mm); 10^{-6} m (1 µm).

(e) Area: (Square metre); 10^4 sq. m. (1 hectare); 10^2 hectare (1 sq. km), 1 hectare (2.471 acres), 1 sq. m = 10^2 sq. dm = 10^4 sq.cm = 10^6 sq. mm.

(f) Weights: (Gram); 10^3 g (1 kg), 10^3 kg (1 Mg, Megagram); 1 Mg (1 Tonne);
(Actually mass) 1 Tonne (10 quintal); 1 g (5 Metric carat), 1 carat (200 mg), 1 g = 10^3 mg = 10^6 µg.

(g) Cube: 10^3 cubic mm = 1 cubic cm; 10^3 cubic cm = 1 cubic dm; 10^3 cubic dm = 1 cubic metre.

1 cc = 1 ml; 10 ml = 1 cL; 10^3 ml = 1 L = 10^2 cL.

10^2 L = 1 hL; 10^3 L = 1 kL = 10 hL.

[B] Miscellaneous measurements:

(a) Imperial measurements:

(i) Length: 1 inch = 10^3 milli inch, 12 inches = 1 foot, 3 feet = 1 yard; 5.5 yards = 1 rod; 220 yards = 1 furlong = 40 rods; 5280 feet = 1 mile = 1760 yards = 8 furlongs, 3 miles = 1 league; 1 rod = 25 links; 100 links = 1 chain, = 66 feet = 22 yards = 4 rods.

(ii) Area: 44 sq. inches = 1 sq. foot; 9 sq. feet = 1 sq. yard; 30.25 sq. yard = 1 sq. rod; 484 sq. yards = 1 sq. chain; 1210 sq. yards = 1 sq. rod; 4840 sq. yards = 1 acre = 160 sq. rods; 640 acres = 1 sq. mile.

(iii) Volume: 1728 cubic inches = 1 cubic foot; 27 cubic feet = 1 cubic yard = 46656 cubic inches.

8 fluid drachms = 1 fluid ounce; 5 fluid ounce = 1 gill, 20 fluid ounce = 1 pint; 40 fluid ounce = 1 quart = 2 pints; 160 fluid ounce = 1 gallon = 8 pints = 4 quarts; 2 gallons = 1 peck; 4 pecks; = 1 bushel; = 8 gallons; 64 gallons = 1 quarter.

(iv) Weight: 537.5 grains = 1 ounce = 16 drams; 7000 grains = 1 pound = 256 drams = 16 ounce; 14 pounds = 1 stone; 2 stones = 1 quarter; 100 pounds = 1 cental; 112 pounds =1 cw (hundred weight) = 4 quarters = 8 stones, 20 cwt =1 ton = 2240 pounds = 35840 ounce.

(b) Nautical measurements:

(i) 6 feet = 1 fathom; 100 fathoms = 1 cable; 10 cables = 1 nautical miles = 6000 feet.

(c) Ayurvedic measurements:

(ii) 1 rati = 125 mg; 8 rati = 1 g; 6 rati = 1 anna; 12 rati = 1 masha; = 1.5 g = 2 annas; 4 annas = 1 siki = 3 g = 24 rati.

List of Helpful Books

[A] PRACTICAL ORGANIC CHEMISTRY

1. Agarwala, Advanced Practical Organic Chemistry.
2. Clarke, A Handbook of Organic Analysis—Qualitative and Quantitative.
3. Fieser and Fieser, Reagents in Organic Chemistry.
4. Mann and Saunders, Practical Organic Chemistry.
5. Huntress and Mulliken, Practical Organic Analysis.
6. Middleton, Systematic Qualitative Organic Analysis.
7. Visnoi, Advanced Practical Organic Chemistry.
8. Vogel, A Textbook of Organic Preparations.
9. Vogel, Elementary Practical Organic Chemistry (Part 3, Quantitative Organic Analysis).
10. Das, Advanced Practical Chemistry.
11. Kemp, Practical Organic Chemistry.
12. Sengupta, Prayognistha Rasayan (in Bengali)
13. Owen, Characterisation of Organic Compounds by Chemical Methods.
14. Vogel, Practical Organic Chemistry.

[B] REFERENCE BOOKS

1. Brown, Hydroboration.
2. Gaylord, Reduction with Complex Metal Hydrides.
3. Hodgman, Ed. Handbook of Chemistry and Physics (America).
4. Murray and Williams, Organic Syntheses with Isotopes.
5. Olah, Friedel–Crafts and Related Reactions.

6. Thomas, Deuterium Labelling in Organic Chemistry.
7. Theilheimer, Synthetic Methods in Organic Chemistry.
8. Weiberg, Oxidation in Organic Chemistry.

[C] SPECTROSCOPY

(1) Gillam and Stern, Introduction to Electronic Absorption Spectroscopy in Organic Chemistry.
(2) Jaffé and Orchin, Theory and Application of Ultraviolet Spectroscopy.
(3) Scott, Introduction of the Ultraviolet Spectra of Natural Products.
(4) U.V. Atlas of Organic Compounds (Butterworth).
(5) Cross, Introduction to Practical Infrared Spectroscopy.
(6) Bellamy, The Infrared Spectra of Complex Molecules.
(7) Nakanishi, Infrared Absorption Spectroscopy.
(8) Sadtler Catalogue, Philadelphia
(9) Dyer, Application of Absorption Spectroscopy of Organic Compounds.
(10) Bhacca and Williams, Application of NMR Spectroscopy in Organic Chemistry.
(11) Jackmann, Application of Nuclear Magnetic Resonance Spectroscopy in Organic Chemistry.
(12) Roberts, Nuclear Magnetic Resonance, Application to Organic Chemistry.
(13) Silverstein and Bassler, Spectrometric Identification of Organic Compounds.
(14) Biemann, Mass Spectrometry; Organic Chemical Applications.
(15) Budzikiewicz, Djerassi, and Williams, Mass Spectrometry of Organic Compounds.
(16) Hill, Introduction to Mass Spectrometry.
(17) Dyke, Floyd, Sainsbury and Theobald, Organic Spectroscopy, An Introduction.
(18) Beynon, Mass Spectrometry, Application to Organic Chemistry.
(19) Bible, Interpretation of NMR Spectra.
(20) Hamming and Foster, Interpretation by Mass Spectra of Organic Compounds.
(21) Sklar, Mass Spectrometry of Natural Products.

INDEX